GIS for Ecology: An Introduction

GIS for Ecology: An Introduction

Richard Wadsworth
Institute of Terrestrial Ecology, Monks Wood

and

Jo Treweek
Institute of Terrestrial Ecology, Monks Wood

Longman

Addison Wesley Longman Limited
Edinburgh Gate, Harlow
Essex CM20 2JE
England

and Associated Companies throughout the World

First published 1999

ISBN 0 582 246 52 0

British Library Cataloguing-in-Publication Data
A catalogue record for this book is
available from the British Library.

Library of Congress Cataloging-in-Publication Data
A catalog entry for this title is
available from the Library of Congress.

Set by 35 in 11/12pt Adobe Garamond
Produced by Addison Wesley Longman Singapore (Pte) Ltd.,
Printed in Singapore

Dedication

This book is dedicated to our partners and families in memory of many lost evenings and weekends (now where have they read that before!)

Contents

Contents

Preface

In this book we hope to introduce you to some of the ways in which geographic information systems (GIS) can help in your ecological and environmental studies. There is nothing particularly mysterious about GIS although given the density and complexity of some of the jargon you might be forgiven for thinking that there is. As far as possible we try and avoid too much jargon but where the alternative would be tediously repetitive we introduce it with appropriate glossaries. Although GIS are computer-based you do not need to be a computer whiz to use one, nor do you have to be particularly numerate. We make several references to features that can be found on the Internet and you may find access to the World Wide Web useful but not essential.

To some extent this is an idiosyncratic view of GIS; certainly we believe that there are only two reasons to become involved with GIS:

- it lets you perform a task better than you could manually, or
- there is no other practical way to perform the task.

The purpose of this book is to help you decide if your project, task or study falls into one of those two cases. There are (we hope) plenty of reasons to use GIS throughout this book, but many boil down to a question of *reliability*: being able to combine different data sets reliably, to make reliable measurements, to make inferences using reliable data; or of *consistency*: consistent classifications, consistent representation, consistent results. We try and illustrate the techniques we have found useful with practical examples using three contrasting GIS systems; however, given that several hundred different GIS packages exist we offer concepts and ideas rather than specific 'recipes'.

A GIS is nothing until it is populated with appropriate data. Getting hold of the data, understanding the error and uncertainties in the data, importing it into the GIS and appreciating the context in which it was captured are central concerns of this book. We are also excited by novel and increasingly affordable ways to capture data, especially remote sensing and global positioning systems and the potential of GIS to organise and illuminate this data.

We believe GIS can help bridge the gap between theoretical and applied aspects in ecology, especially through their role in:

- visualisation and communication (helping people to understand what is known about the distribution of a particular species or phenomenon);

- audit and inventory (helping people to understand and measure how much of a resource is present and how it might be changing); and
- analyses, prediction, modelling and decision making (helping people to understand the significance of information and to make better-informed decisions).

Structure of the book

Chapter 1 introduces and defines what geographic information systems and outlines some of the types of ecological questions that they can assist with.

Chapter 2 defines just what is meant by spatial data and how it can be represented and manipulated within a GIS.

Chapter 3 tells you how to get started, and how you are going to choose what software and hardware you need.

Chapter 4 will describe some of the many sources of data and the ways in which the data can be captured.

Chapter 5 introduces the basic functionality of GIS, especially for measuring, searching, interpolating and modelling spatial data.

Chapter 6 indicates how, having collected your data, imported it in the GIS and done some analysis, you now need to visualise and communicate your results.

Chapter 7 provides a series of case studies that we hope will inspire you to take the plunge and start to use GIS in your studies of ecological processes.

Acknowledgements

We would like to thank Neil Veitch for his valiant attempt to get this book started before he moved onto other projects. We also wish to acknowledge the thoughtful and constructive comments of our colleagues on the later drafts of this work, who it would be invidious to attempt to identify individually.

We wish to thank the Institute of Terrestrial Ecology for permission to reproduce as Plate 4 an image of the Land Cover Map of Great Britain, and Oxford University Press for permission to reproduce an extract from *Managing GIS Projects* (Huxhold and Levinsohn, 1995). We also wish to thank the authors and Kluwer academic publishers for permission to reproduce Figure 7.1 (from *Environmental Monitoring and Assessment*) and Figure 7.3 (from *Landscape Ecology*) also, thanks to Blackwell Science and the authors for Figure 7.2 (from the *Journal of Applied Ecology*), and Jonathan Cooper for supplying the albatros image for the front cover. Full references are given in the text.

Chapter 1

Geographic information systems, ecology and the environment

Introduction

Ecology is the study of how living organisms, including man, interact with their environment. In addition to the pure intellectual challenge of understanding ecological systems there are many pressing environmental problems facing the world today that demand better understanding of our own ecological role and impact. If the growth and development of society and civilisation is to be sustained without irreversible damage to the ecosystems on which people depend, then it is imperative that the consequences of natural resource exploitation and management by people should be understood. Fortunately, there is increasing awareness of environmental problems and at least some political will to do something about them.

The widespread introduction of environmental impact assessment (EIA) is one example of measures that have been implemented specifically to address the environmental consequences of human economic activity. International co-operation to tackle environmental problems has also developed, as exemplified by the Earth Summit in Rio in 1992 and by various international conventions on pollution, the protection of wetlands (RAMSAR: i.e. the Ramsar convention on wetlands of international importance held in Ramsar, Iran, 1971), trade in endangered species (CITES: i.e. the Washington Convention on the International Trade in Endangered Species of flaura and fauna) and so on. The important thing to remember is that 'good' decision making depends in part on 'good' **information** and the ability to interpret it. Because of the complexity of environmental systems, we tend to study them using a reductionist approach, focusing on small, discrete, simplified aspects. However, most environmental problems are multi-faceted and demand consideration of a diversity of information, issues and interests. This is where geographic information systems (**GIS**) come in. As we hope to demonstrate, GIS populated with the right data and models can help in the organisation, interpretation and communication of ecological information in an efficient and effective manner.

A GIS is a computer-based system to input, store, manipulate, analyse and output spatially referenced data. A GIS combines data, **hardware**, **software** and **liveware** (which is you!). This of course means that you may have to learn

1

several sets of jargon for computer technology, programming and software, **spatial analysis**, **cartography** and (possibly) some perceptual jargon.

At the heart of a GIS is a database that allows the linkages between spatial data and attribute data to be made. Hardware might include equipment to capture information (**scanners**, **digitisers** and so on), store and archive spatial data (various disk and tape drives) and to output the transformed and manipulated information (screens, printers, plotters and so on). Software provides the means to manipulate and analyse spatial data. Liveware is *you*, the person who has to decide what analysis to carry out, how it should be performed and why.

The specialist literature on GIS can be extremely technical and is often impenetrable to those with no specific training in the field. GIS can encompass aspects of software engineering and systems design, spatial statistics and mathematical modelling, perceptual psychology and traditional cartographic design, and it is almost impossible to become completely familiar with all aspects of it. For most ecologists, there is much more to be gained from becoming *spatially literate*, knowing what GIS has to offer and being able to decide when and how it should be used to tackle ecological problems. As applied research scientists, we became involved with GIS because there was no other sensible way to tackle the particular issues and problems we were interested in. This book is intended to provide the basic advice, information and encouragement we would have appreciated when we first set out to use GIS for ecological study and application. The approach is deliberately pragmatic and may seem over-simplistic to the more experienced GIS user. There is a wealth of literature on the more technical aspects of GIS development and application, and where this is both relevant and likely to be reasonably accessible, we try to point the interested reader in its direction.

The best way to learn about GIS is to start using it. The initial learning curve will be steep (we can still remember the satisfaction of getting our first map on the screen in the right place with the right colours) and the GIS jargon may seem obscure and daunting at first, but do not be deterred. It should not take long for your confidence and skills to grow to the point where you can decide what approach to take and what tools you will need to solve a particular problem.

Structure of this book

As you are a busy person we have tried to structure this book so that you can rip out the information you require as quickly, reliably and efficiently as possible. Boxes and exhibits are liberally scattered throughout the text highlighting critical points, jargon and terminology, advice and warnings. We start by defining GIS and outlining the history of its development. We then launch into the thorny subject of what spatial data are and how to get hold of such data before going on to discuss some of the unique manipulations that are possible in a GIS and then how information can be presented. A series of longer case studies are grouped together in the final chapter and are intended to illustrate the different roles that GIS can play in ecological studies. Different analytical techniques are illustrated with relevant ecological examples where possible.

We have attempted to avoid too much jargon in this book but it is sometimes more efficient to use a specific word rather than spelling out the meaning every time. A glossary of terms is provided at the back of the book, but we try to **highlight** and explain new terms as they are introduced.

The origins and history of GIS

We are familiar with receiving much of our spatial information about the world from maps. In fact, until relatively recently, the 'paper' map, sometimes with a memoir, was the only form of spatial database that existed. Maps have been around since the dawn of civilisation and GIS with its roots in mapping technology has a surprisingly long history. Box 1.1 provides some significant dates (mostly from Coppock and Rhind, 1991).

A number of limitations are shared by any map produced on a permanent medium, whether it is a Babylonian clay tablet or the latest 'high-tech' glossy printer paper (Box 1.2). Developments in computer technology in the 1960s and 1970s made it possible to automate the mapping process (digital cartography), thereby removing some of these limitations. As well as making it quicker, easier and cheaper to produce and update maps, additional benefits of automation soon became apparent. These included the ability to create maps for specific users, experiment with different graphic representations, facilitate interactions between statistical analyses and mapping and more easily control the effects of classification and generalisation on the quality of input data.

While GIS evolved out of the application of computers to cartography, GIS and digital cartography should not be considered identical. A GIS does not actually contain maps: rather it contains the information from which maps can be generated. A GIS can also be distinguished from database management systems or from visualisation packages through its specialised capability for **spatial analysis** (Albrecht, 1996, provides a rigorous discussion).

As well as new ways to produce maps, new ways to collect and process data have appeared over the last few decades. These include satellites, which provide synoptic information over very large areas at frequent intervals, and global positioning systems (**GPS**), which allow for the collection of point data without the need for an extensive and expensive ground control network. For ecologists, the availability of new sources of data, combined with the GIS technology needed to store, interpret and display it, has opened up whole new areas of research and application.

Box 1.1   Some significant dates in the history of geographic information (mostly from Coppock and Rhind, 1991)

1137   Earliest known map with a regular grid (it was carved on a stone tablet in China with a grid spacing of approximately 100 li).

1569   A major breakthrough in the production of maps for quantitative uses when Mercator publishes his map as a diagram to aid navigation.

1958   David Bickmore identifies potential of computers for cost-effective editing and classification of cartographic information.

1960   Geographic Data Technology Inc. founded.

1962   *Atlas of British Flora* produced: all the maps were produced in an automated manner by modifying punch card tabulators.

1963   Harvard Laboratory for Computer Graphics starts work on a computer mapping system.

1964   SYMAP produced: it is able to construct **isoline, choropeth** and proximal maps using line printers.

1966   First version of the Canadian Geographic Information System (CGIS) available (although the system was not fully operational until 1971).

1968   First GIS conference organisers are the International Geographic Union's Commission on Geographic Data Sensing and Processing.

1969   The Environmental Systems Research Institute (ESRI) is founded, it produces Automap II in 1970 and ArcInfo in 1980.

1969   Intergraph founded.

1970   University of Oregon produces the Polygon Intersection and Overlay System (PIOS).

1970   The Experimental Cartographic Unit (ECU) publishes first digital map in the UK (geology of Bidston).

1973   Handling of large data sets is helped by the introduction of the Geographic Information Retrieval and Analysis System (GIRAS).

1973   The Ordnance Survey (the national mapping agency of Great Britain) begins digital map production, complete digital coverage at least at 1 : 10 000 **scale** achieved within 15 years.

1975   Harvard Laboratory produce POLYVRT.

1977   First 'modern' GIS, ODYSSEY, produced; full working version completed in 1979.

1980   First version of ESRI's ArcInfo released.

1987   Microsoft's first GIS, Map-Info, released.

1987   Clark University releases first version of Idrisi.

Box 1.2   Limitations of 'paper' maps

- Original data are greatly reduced in volume and usually reclassified in order to make them understandable and visual: this process often filters out many local and possibly crucial details.
- A map has to be drawn extremely accurately. Standard cartographic practice is to produce line detail to within fractions of a millimetre and the presentation needs to be very clear.
- Several map sheets are often required to cover an area making a synoptic view difficult.
- Once data are displayed on a map it is not cheap or easy to retrieve them in order to combine with other data.
- A printed map is static and it is extremely difficult to attempt quantitative spatial analysis without collecting new information.
- While conventions exist for the representation of surfaces on printed maps the representation of volumes is much more difficult.
- Typically a printed map represents the conditions at a particular instant of time: time series stochastic and cyclical variations are difficult to portray.

What is a geographic information system?

The physical (or hardware) components of a GIS are shown in Figure 1.1. Note that with the advent of computer networks not all systems will necessarily have all their components in the same physical location. The GIS on your desk might consist physically of just a keyboard and visual display unit (VDU, i.e. the computer screen), other equipment such as a hard disk (to store data and software), a central processing unit (CPU), compact disc (CD) readers and writers, printers, plotters, and more specialised equipment such as scanners and digitisers can be located elsewhere and accessed through a local or wide area network (LAN or WAN).

The relatively sophisticated software products that are available today provide tools that help the user to capture, store, manage, manipulate, analyse and display spatial data. Typically GIS software has five components:

- tools to help import data,
- a database,
- a database management system,
- tools to transform and analyse spatial data, and
- tools for displaying and printing data.

GIS software differs from most other types of software because of the need for 'real' input data (compare this with say, a word processor, where you make up the 'data' as you go along!). Obtaining the input data for a GIS is often a long, hard, expensive operation. Understanding the data and making sure the database is fit for its intended purpose may take even longer. Data models (abstractions of reality) are covered in Chapter 2 and sources of data in Chapter 4.

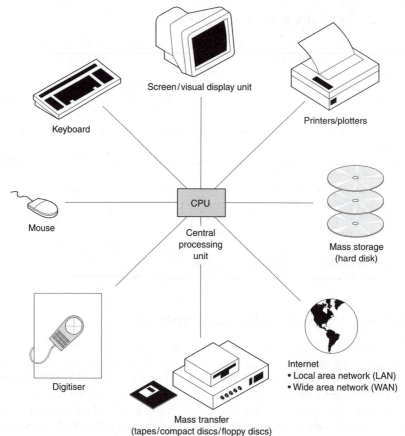

Figure 1.1   Basic physical components of a GIS: the hardware.

GIS software performs various functions; Huxhold and Levinsohn (1995) suggest three primary functions:

- automated mapping functions,
- data management functions, and
- spatial analysis functions,

with additional support functions for:

- communications (specifically electronic file transfer and networks);
- menu and graphical user interface **(GUI)** design;
- creation of symbols and other cartographic design features;
- interfaces with existing packages; and
- customised programming, particularly file archiving and batch processing.

At the moment there are several hundred GIS software packages available. There is almost certainly a GIS that will meet your needs, run on the computer hardware you already have and be within your budget (advice on choosing what you need is given in Chapter 3 and examples from three software packages are shown in Chapter 5).

A GIS can be used be used to study phenomena ranging from those that operate over a few millimetres (such as competition between plants in a grass sward) to those that operate over many kilometres (say migration of birds or mammals). This raises the thorny issue of what is meant by **scale**: colloquially large and small scale refer to size of the study area; cartographically scale refers to the ratio of the size of the site to the size of the map. Study of leaves in a sward would be colloquially a *very small-scale* study, cartographically it is a *very large-scale* study; conversely a study of geese migration is colloquially a *large-scale* study but cartographically it is *very small-scale*. Where possible we will try to be specific about what we mean when we refer to scale, but if in doubt assume we are using scale in its cartographic sense. Consideration of scale (and how to scale up and down) forms a major area of research in several areas of the natural sciences (especially hydrology and ecology) and for a rigorous treatment an interested reader is referred to Gardingen *et al.* (1997) or Wiens (1989).

Typically a GIS is used to represent space in two dimensions in the same way as conventional maps. Three- and four-dimension representations are being developed and for obvious reasons marine applications of GIS have taken the lead in 'higher' dimension representations. It may also be appropriate to conceptualise a forest or a soil pedon as a three-dimensional structure within which an organism may move, rather than as a two-dimensional 'map'. Maps are usually seen as the basis of GIS as they provide a convenient 'front end' through which the data, which constitute your model of the world, can be accessed. In fact the key element is the ability to use geo-referencing to tie together (relate) all the information, whatever its source, in a quantitative manner. If data can be tied to a unique known location then just about any sort of information can be incorporated into a system. This might include text, graphs, diagrams, photographs, video footage or sound as well as satellite imagery, aerial photography and ground survey observations. All sorts of information can be included in the system that can help you either *understand* the data or *communicate* your research. Quantitative analysis and manipulation is a much more challenging and interesting use of GIS, to explore not just 'what happened' but 'why did it happen where it did', 'how did it happen' and 'what is it going to do in the future?'.

It is not our intention to provide a 'cook book' or 'user manual' but we hope to introduce you to some of the more interesting things that GIS can do for ecologists.

What use are GIS to an ecologist?

Maps have always performed an important role in ecological study and will no doubt continue to do so. It is therefore worth considering what it is that makes GIS potentially more useful than straightforward paper maps in the study of ecological phenomena. Conventional two-dimensional ('paper') maps typically provide two types of information: 'spatial information' about *where*

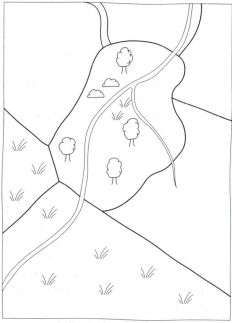

Figure 1.2 A simple sketch map (not drawn to scale, not quantitative, not annotated, but still full of information).

something is, what shape it is and how it is located with respect to other features; and 'attribute information', which conveys *what* the feature is. As a simple illustration consider the woodland on the sketch map (Figure 1.2). Even from a simple, two-dimensional map like this, a wealth of information can be obtained. For example, we can observe that a wood exists in the area described by the map and can draw conclusions about its approximate shape and area. We can also observe that a stream passes through part of it and conclude something about the topography of the area and other land uses. From a two-dimensional representation of a landscape like this one, it is straightforward to extract qualitative (descriptive) information, but much harder to extract quantitative information, particularly so in this example because no scale is given. For most ecological studies we need more spatially explicit and quantified information about the factors that might actually explain the observed distributions and patterns of landscape 'features' like woodlands. By adding information about geology, climate, soil type and the distributions and behaviour of woodland species to our map, we might be much better placed to understand the woodland as a habitat rather than simply an observed grouping of trees. However, it is difficult to see how all this information could be incorporated into the map shown in Figure 1.2 without causing consider-able chaos and confusion. Furthermore, it is important to consider how informa-tion about the locations and distributions of ecological features or processes is transferred 'from the ground' to the paper. There is considerable scope for error as the exact locations (co-ordinates) are unknown.

Having seen how much information is available on a sketch map, why bother with a GIS? There are (we hope) plenty of reasons to use GIS throughout this book, but many boil down to a question of *reliability*: being able to combine different data sets reliably, to make reliable measurements, to make inferences using reliable data, or of *consistency*: consistent classifications, consistent representation, consistent results.

The evolution of GIS has opened up whole new areas of spatial analysis in ecology that would otherwise be too laborious to contemplate.

Most commonly GIS are used to portray landscape features on the Earth's surface, (although topographic data now exists for Mars there is no extra-terrestrial ecological data to go with it yet!). As long as a common spatial co-ordinate system is used for the input data a GIS can be used to represent vertical maps, of say a root system, as well as horizontal systems.

Despite the potential advantages of GIS-based study, developing a GIS can involve considerable investment in time and other resources, so why should you bother? There are only two circumstances that justify commitment to GIS:

- it lets you complete the task better than you could do manually,
  or,
- it lets you perform a task you cannot do any other way.

There is no such thing as a 'GIS project', only 'projects that need GIS'. In other words, do not be led by the technology: use it as a tool.

How are you going to decide if a GIS might be useful to you? Most GIS textbooks start with a series of questions posed in abstract or generic terms. Because we do not want to feel left out we include a typical list of abstract questions about spatial processes (the 'classic' list is in Burrough, 1986). Because it can be difficult to see how a particular practical problem relates to an abstract question we have included a series of more 'applied' questions. If you have ever asked, or wanted to ask, questions of the type(s) listed below (Box 1.3) then you might want to consider learning more about GIS. Any of these questions are difficult or at the very least time-consuming to answer using manual methods, but can usually be answered quickly, simply and consistently if the data are entered into a GIS.

GIS can also be considered in terms of more general roles, including:

- visualisation and communication (helping people to understand what is known about the distribution of a particular species or phenomenon);
- audit and inventory (helping people to understand and measure how much of a resource is present and how it might be changing); and
- analyses, prediction, modelling and decision making (helping people to understand the significance of information and to make better-informed decisions).

As a visualisation and communication tool GIS can help to reveal patterns in data, to detect possible underlying trends, to identify clusters of observations or gaps in data and to speculate about the ecological significance of observed patterns.

Box 1.3   Spatial questions that might indicate a GIS–based approach

'Ecological' form of the question	'Generic' form of the question
• How are the ash trees distributed in my local nature reserve? • Where do the first observations of swallows occur in the spring? • Where are the quadrats recorded by my research group?	Where is object A?
• What is the relationship between prairie and forest as you travel towards the Arctic Circle? • Are man orchids more likely to be found on south-facing or north-facing slopes? • How does the diversity of sphagnum species vary along an east–west gradient across the country?	Where is object A in relation to place B?
• What is the area of this local nature reserve? • I have a series of observations (by radio tagging) of an individual muntjac deer; what is the probable size of its home range? • How does seed production of *Impatiens glandulifera* relate to the area occupied by an individual plant?	How large is A?
• What is the ratio between the length of the boundary and the core area of my local nature reserves? • How long is the river between the two local towns? • How has the boundary of heathland changed since the first published maps?	What shape (two dimensions) is A?
• Where is the hill slope convex or concave? • Where are the steepest slopes and highest risk of erosion? • Where will water accumulate (or floods occur) after heavy rain?	What shape (surface and three-dimensional shapes) is A?
• What proportion of grouse nests are predated within 100 m of the edge of conifer plantations? • Are there fewer than the expected number of lichen species within 10 km of the aluminum smelter? • How many woodlands are there within 200 m of the proposed new trunk road?	How many occurrences of type A are there within distance, D, of place B?
• What is the expected rainfall at my study site? • How does the solar radiation budget vary over the site? • How often is the edge of the salt marsh under water during the year?	What is the value of function Z at position X?

(continued)

(continued)

'Ecological' form of the question	'Generic' form of the question
• What is the relationship between rainfall, temperature and sphagnum species diversity? • How is the frequency of fire related to topography and species composition? • What is the habitat suitability of deciduous woodland on a clay soil at low elevation?	What is the result of intersecting these layers of spatial data?
• What is the shortest 'practical' route for an individual between two badger setts? • What is the least environmentally damaging route for a new road to take? • What is the relationship between 'genetic distance' and 'physical distance' among a population of *Impatiens glandulifera* along a river bank.	What is the path of least cost, resistance or distance between X and Y?
• What vegetation is present at a series of random locations? • What species are present on these islands? • What is along the proposed routes of the new road?	What is at point $X_1, X_2, \ldots$, etc.?
• How does the severity of the outbreak of an insect pest depend on the combination of plant species in the surrounding fields? • What species are present on islands that are close to the centre of the archipelago? • How does the boundary of the home range vary with the perceived habitat suitability for a species?	What features are next to objects having a certain combination of attributes?
• Given some measurements of the **reflectance** of a piece of land, what is the most likely vegetation cover? • What regions have a unique combination of soil, vegetation and topography? • Which regions have similar combinations of soil, vegetation and topography?	How are objects having certain combinations of attributes classified?
• What is the effect on plant species diversity of increased grazing pressure due to changes in livestock subsidies to farmers? • What is the rate of erosion likely to be in this river catchment if the climate continues to be drier than the long-term average? • What is the long-term effect on the ground flora of the cessation of regular coppicing in the woods of lowland Britain?	What is the effect of process, P, over time, T, for a given scenario, S?

As a tool for audit and inventory GIS can help with the management or organisation of spatial data. New information can be added to the GIS as it becomes available or is revised or updated. Some data transformation may be needed before information can be displayed or manipulated; however, it is increasingly possible to store data close to their 'native' form and transform or classify the data as and when required (on the fly). It should be noted that such a data policy will slow down your computer, but it saves you a lot of time and work in justifying and documenting tedious data transformation. The measurement facilities in the GIS are particularly useful for carrying out inventories and audits by determining how much (area, length, shape, counts, distances, etc.) of a particular resource exists now.

It is with respect to analysis, prediction and modelling that GIS transcends other software packages. Most GIS contain facilities to help with the statistical analysis of data and many incorporate models of various types.

Many who have seen the colourful and attractive slides that accompany talks on GIS will wonder if they represent more than an expensive and elaborate (and possibly time-consuming) way to produce pretty pictures. But as Johnston (1990) points out, 'Its power for landscape ecology lies in its ability to manipulate and analyse spatially distributed data. While community and ecosystem ecology focused on the central tendency of observations made at carefully selected points in space, landscape ecology focuses on the heterogeneous distribution of ecological resources, populations and processes over time.'

What you can use instead of a GIS

We do not want to claim that a GIS will always be the answer to all your problems. It may be that the analysis you need to do can be done just as well, or even better, using a different approach. You might be able to 'force' a spreadsheet, statistical package or **relational database** to do the spatial analysis you need. But if you can assemble the data on a computer and master the complexity of a spreadsheet or database we see no good reason not to use a GIS. The principal alternatives to GIS you might need to consider are relational databases, computer-aided design/computer-aided management (**CAD/CAM**), programming languages, spreadsheets, image analysis, drawing packages and statistical packages.

Databases

In fact databases are usually integral to GIS; many are linked to a particular database, but increasingly GIS software will link to a variety of different packages and sometimes to several simultaneously. Many of the spatial analyses carried out by a GIS can be carried out using a database (although phrasing the questions is somewhat intricate). If you are already fluent with a database package, you do not have too much spatial analysis and you can manage with limited graphical output then a database may be sufficient.

Digital cartography

This should be viewed as a separate activity from a GIS (although of course most GIS include some aspects of digital cartography). If you have a requirement to produce very high-quality maps rather than to carry out spatial analysis, then a digital cartography package may be preferable to a GIS.

Computer-aided design/computer-aided management (CAD/CAM)

A CAD/CAM package will contain many powerful display and design features. If your interest lies in very small study sites (large cartographic scales) then it may be more appropriate.

Programming languages

In the 1980s a standard called the graphical kernel system (**GKS**) was produced that allowed graphics to be produced using the Fortran programming language. Fortran programs and simple databases were used by several ecological organisations (including the one we now work for) before GIS became widely available. More recently Visual Basic and C++ have included facilities for producing graphical output.

Spreadsheets

For relatively small **raster** data sets spreadsheets can be very convenient; however, they tend to slow down very quickly as the data set gets larger. This book uses a spreadsheet to illustrate some aspects of the display and manipulation of spatial data.

Image analysis

Many image analysis packages for remote sensing are converging in functionality with GIS, but they still maintain a separate identity. If your main priority is the classification and display of remotely sensed data then an image analysis package may be sufficient (or preferable).

Drawing packages

There are many different drawing packages; however, they always work in terms of page co-ordinates. They are very useful in making the output from a GIS up to 'cartographic' standard, but this can be very time-consuming.

Statistical packages

Most statistical packages contain plotting utilities. Some, such as Uniras, contain specific mapping routines (Unimap).

Review and preview

In this chapter we have tried to provide a working definition of GIS and a little of the historical background to its development. More importantly we have tried to introduce you to what we believe is its potential to help answer environmental and ecology questions. Over the following chapters we will try and elaborate on these themes as well as introducing several more. In Chapter 2 we define exactly what we mean by *spatial data* and some of the practicalities and problems of how spatial data are perceived and represented. In Chapter 3 we provide some guidance on starting your own GIS project, especially the importance of deciding exactly what you are trying to do. A GIS is nothing without data and in Chapter 4 we describe some sources of data and traditional and novel methods of data capture. In Chapter 5 we get to grips with spatial analysis and in Chapter 6 give some guidance in how to present the results of your analysis. We finish in Chapter 7 with a series of case studies illustrating how GIS has been used in a variety of roles.

Further reading

The bookshelves of most GIS users we know are graced by a well-thumbed copy of P.A. Burrough's *Principles of Geographic Information Systems for Land Resource Assessment*, Monographs on Soil Resources, Survey No. 12 (Oxford Scientific, 1986). The two volumes edited by D.J. Maguire, M.F. Goodchild and D.W. Rhind, *Geographic Information Systems*, (Longman, Harlow, 1991) provide a comprehensive view of GIS at the beginning of the 1990s; and a collection of papers on the environmental applications of GIS is given in J.G. Lyon and J. McCarthy, *Wetland and Environmental Applications of GIS* (Lewis, New York, 1995). Advances in GIS methods and novel applications can be found in the *International Journal of Geographic Information Science* (previously *Systems*) while the *Cartographic Journal* is more concerned with the interpretation and display of spatial data and *Mapping Awareness* provides a professional rather than academic overview of current events. Applications of GIS to ecological problems are found in many journals but we try and check the contents of *Applied Ecology*, *Landscape Ecology* and *Environmental Planning and Management*. Remote sensing journals like *Photogrammetric Engineering and Remote Sensing*, *Remote Sensing of Environment* and the *International Journal of Remote Sensing* regularly included articles on detecting and classifying vegetation cover and properties like biomass. Details of many GIS systems and example applications can be found on the Internet – some specific Web pages are given in Chapter 4 (data) and Chapter 5 (software).

Chapter 2

Thinking spatially – space, place and time

Spatial data

Every ecological event has a spatial and temporal dimension, in that it occurs at a particular location and at a particular time. Ecologists need to know where and when observations were made and what was recorded to make sense of them and draw conclusions about ecological processes. To make generalisations about or from sets of observations, the relationship between those observations and others is important. GIS handle data in such a way that the distributions of ecological observations in space and time can be quantified and more easily characterised and understood.

GIS are concerned with two types of data:

- location (typically expressed as: x–y, or eastings–northings, or latitude–longitude co-ordinates); and
- attributes (type of observation, counts, timing, size, shape, etc).

Sets of attributes, together with their recorded locations, may be stored, classified and represented in various ways within the GIS. For example, related attribute data might be selected and grouped to constitute a theme, like 'soils', 'rainfall' or 'vegetation'.

Location

The location of an object or feature is defined within a co-ordinate system. A co-ordinate system consists usually of two reference lines crossing at right angles (orthogonally) but may consist of a reference point and a reference direction. The most common co-ordinate systems for spatial data are the **Cartesian co-ordinate** system (for use on a flat surface) and the latitude–longitude system (for use on the globe).

Under the Cartesian system the point where the axes (reference lines) cross is termed the **origin**. Locations are defined by a co-ordinate pair that represents the perpendicular distance from the origin along each **axis**. This process is illustrated in Figure 2.1. Co-ordinates are typically expressed in terms of x–y or easting–northing. The rule is that the x co-ordinate or easting is written first then the northing (one way to help you remember is the phrase:

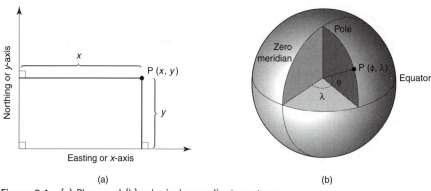

(a)   (b)

Figure 2.1   (a) Plane and (b) spherical co-ordinate systems.

'in the door and up the steps'). Note that when the origin is centrally located co-ordinates can be positive and negative. Because this can be confusing it is common to add a large constant number so that co-ordinates are always positive, creating a **false origin**. For example in the UK, rather than use the true origin, 400 000.0 m is added to all eastings values, which puts the false origin well out in the Atlantic Ocean.

For co-ordinates on a sphere the two reference lines are *great circles* and must therefore cross at two points (if you cut a sphere through the centre the cut surface is a great circle). Typically one reference line is the Equator; the second is an arbitrarily selected line of longitude. Latitude (ϕ) is measured from zero at the Equator to plus 90° at the North Pole and to minus 90° at the South Pole. Longitude (λ) is measured from the zero meridian, positive westwards and negative eastwards. This is illustrated in Figure 2.1. Note the conventional order is to show latitude first and then longitude, which is the opposite way around from the Cartesian co-ordinate system. (If you have trouble remembering which is latitude and which is longitude – lines of *long*itude are the ones which are all the same *length*).

Note that how angles are 'used' in a GIS can be confusing. Mathematicians measure angles from the *x*-axis with positive increase counter clockwise. Surveyors and geographers, on the other hand, measure bearings (angles) from due north (the *y*-axis) with bearings increasing in a clockwise manner. You may need to discover whether your GIS uses the mathematicians' convention or the geographers' convention.

Map projections

A map **projection** is a set of *mathematical equations* used to convert between co-ordinates in latitude, longitude and co-ordinates in eastings, northings and vice versa. There are many different map projections. The problem with representing a sphere (or **spheroid**) on a flat piece of paper is well known. This problem is usually expressed in terms of trying to lay the peel of an orange flat. As an alternative try drawing a large triangle on your orange and measuring the angles. As you will remember on a plane the internal angles of a triangle must add up to 180°, on a sphere (orange) they must add up to more than 180°. The importance of this example is that it demonstrates the absolute impossibility of maintaining all the geometric properties on a sphere that exist on a plane. A map projection may preserve area, shape (at a point) or distance but it cannot preserve them all (and some projections preserve none of them). Some map projections can be visualised in terms of how a piece of paper could be wrapped around a globe to form a cylinder or cone or just laid flat touching at one point. Other projections can only be represented mathematically. National mapping agencies generally choose conformal projections, which maintain correct angles at a point so that angles you measure on the ground (with a **theodolite** say) are the same as the angles on the map *at that point.* An alternative property that is useful for global displays is that the relationships between the size of areas is maintained (termed equal-area projections). The important point to remember is that all map projections introduce some form of distortion and there is no universally good projection; audience, use, tradition and aesthetics all have a role to play in selecting an appropriate projection.

With respect to data you want to use in a GIS, the critical idea is that data are referenced to a known co-ordinate and projection system, i.e. you know which projection was used (and its parameters) and the values used for the shape of the Earth.

In 1569 Mercator produced his first maps specifically to help mariners get from one place to another, and for navigation his projection (usually in its transverse case) is still a widely used conformal projection. Unfortunately the projection has been used to produce maps of the whole Earth – which is something Mercator himself never did – the inappropriateness of the projection for such a task has generated a considerable literature. There are many equal-area projections that generate a rectangular image of the world, so many in fact that a projection published in 1855 (by Gall) needed to be 'rediscovered' by Arno Peters in 1967 (see Dorling and Fairburn, 1997, for a history of the controversy). Many projections generate non-rectangular images of the world, one of the earliest being the equal-area sinusoidal (devised in 1606); other non-rectangular projections include Mollweide's (devised in 1805) and Robinson's (devised in 1963) but our own all time favourite is Hammer–Aitoff's (Figure 2.2). The simplified forms of the equations used to produce Figure 2.2 are given in Table 2.1 (the simplification is to assume that the Earth is a perfect sphere of unit radius). The equations assume ϕ and λ (latitude and longitude) are given in radians.

Most GIS include many map projection algorithms and spheroid parameters, for example ArcInfo® (version 7) includes 46 projections and 30 spheroids, so some care needs to be taken to use the correct spheroid with the right projection for a particular country. Box 2.1 provides more details on the geoid, spheroid and shape of the Earth. There are many texts on map projections: those that go into the geodetic complexity of projections on triaxial

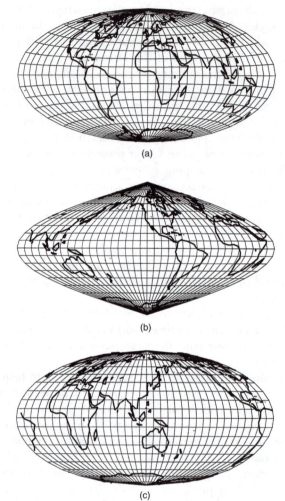

(a)

(b)

(c)

Figure 2.2 'Nice' (non-rectangular) equal-area maps: (a) Hammer–Altoff, (b) Sinusoidal, (c) Mollweide.

Table 2.1 Examples of three equal-area projections

Projection	x co-ordinate	y co-ordinate
Hammer–Aitoff's	$\dfrac{2\sqrt{2}\,(\cos\phi\sin\lambda/2)}{\sqrt{(1+\cos\phi\cos\lambda/2)}}$	$\dfrac{\sqrt{2}\,(\sin\phi)}{\sqrt{(1+\cos\phi\cos\lambda/2)}}$
Mollweide	$\dfrac{\sqrt{8}}{\pi}\big[\lambda(\cos\psi)\big]$ (where $2\psi + \sin\psi = \pi\sin\phi$)	$\sqrt{2}\,(\sin\psi)$
Sinusoidal	$\lambda(\cos\phi)$	ϕ

Box 2.1 Digression on the geoid, spheroid and the shape of the Earth

The terms geoid, spheroid and projection often cause confusion. The Earth is an irregular lump of rock spinning around in space. At this instance there are two forces acting on you: one due to gravity towards the centre of mass of the Earth, the other due to the spinning motion of the Earth (this force has a maximum value at the Equator and a negligible value at the North and South Poles). It is possible to define an equi-potential surface such that everywhere on the surface the magnitude of the resultant of the two forces is the same. The term **geoid** refers to the theoretical equi-potential surface that most closely corresponds to global mean sea level. Because the Earth is irregular the geoid itself is irregular and is not amenable to mathematical treatment. A **spheroid** is an approximation (usually an ellipse rotated around the minor axis) to the geoid. Over the years many spheroids have been defined that give good approximations to the geoid for a particular country or a continent. Latitude and longitude are measured in relation to the geoid because the observed vertical (say a plumb bob) hangs straight down perpendicular to the geoid, so you need to correct for the difference between geoid and spheroid with observed latitude and longitude. A map *projection* is a set of equations that converts from co-ordinates on the spheroid to a set of co-ordinates on a plane surface.

The relation between the Earth's surface, geoid and spheroid are shown in Figure 2.3.

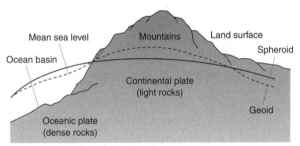

Figure 2.3   Difference between geoid and spheroid.

ellipses are rather daunting; at the other end of the 'market' are 'coffee table' books that stress the aesthetic quality of maps. Our own favourite is a school text that provides instructions for the graphical construction of projections, Steers (1927 – but there was a 15th edition in 1970), while Maling (1972) is more recent.

Several of the major geographical and cartographic institutes and organisations have resolved that rectangular maps (images) of the whole globe are a bad idea. Box 2.2 gives the text of the resolution that they passed (Anon., 1989).

Box 2.2   Good reasons for using really nice projections for whole Earth maps (and not those boring rectangular ones)

WHEREAS, the Earth is round with a co-ordinate system composed entirely of circles, and

WHEREAS flat world maps are more useful than globe maps, but flattening the globe surface necessarily greatly changes the appearance of the Earth's features and co-ordinate system, and

WHEREAS world maps have a powerful and lasting effect on people's impression of the shape and sizes of lands and seas, their arrangement, and the nature of the co-ordinate system, and

WHEREAS frequently seeing a greatly distorted map tends to make it 'look right',

THEREFORE we strongly urge book and map publishers, the media and government agencies to cease using rectangular world maps for general purpose or artistic displays. Such maps promote serious, erroneous conceptions by severely distorting large sections of the world, by showing the round Earth as having straight edges and sharp corners, by representing most distances and direct routes incorrectly, and by portraying the circular co-ordinate system as a square grid. The most widely displayed rectangular world map is the Mercator (in fact a navigational diagram devised for nautical charts), but other rectangular world maps proposed as replacements for the Mercator also display a greatly distorted image of the spherical Earth.

   Resolution passed by:

- American Cartographic Association
- American Geographic Society
- Association of American Geographers
- Canadian Cartographic Association
- National Council for Geographic Education
- National Geographic Society Special Libraries Association, Geography and Map Division

Attribute data

Attribute data can be anything: a name, number, classification or description (which could be verbal, pictorial or numerical). On a conventional map attribute data are contained in the **legend** or *key*, which explains what the map symbols actually mean. For example, woodland on a map might appear as just a green blob until you look at the legend and find out that the 'green blob' is intended to represent woodland. The difference between a conventional map and a GIS is the speed and amount of attribute data that can be retrieved and manipulated for a specific location. Suppose you are particularly interested in 'green blobs' then it should be possible to produce a map (or layer on the GIS) that has the 'green blobs' as the **base map** overlain with the information (additional attribute data) about species, areas, designations or whatever else you may have. Alternatively you may wish to find out how different attributes are related to each other. For example, most of your 'green blobs' might contain the attribute that species X is present and you want to know where else species X is present.

Attribute data for a GIS are usually stored in tabular form in a database that is linked in some way to the stored spatial data. Many GIS use some form of **relational database** for storing attribute data. As the simplest case suppose that you have collected species data from some quadrats scattered at random across a farm. The spatial data are the locations of the quadrats; associated with each point (location) is a unique identifier. In the relational database you need to construct a table where each row holds the data for a single quadrat and each column represents an individual species. Box 2.3 shows the types of question that can now be asked of this simplest possible case (a database consisting of a single table).

The real value of a relational database starts to become apparent as extra 'tables' are added. Information in the tables can be *related* either directly or through other tables because they have some attribute in common. For example, a table might be constructed where the rows are individual species and the columns represent conservation status. Hence questions can be asked about the conservation 'value' of the quadrats (because the first table specifies the quadrat to species, and the second table specifies species to status). As more and more tables are added, more and more elaborate 'chains' of relationships can be constructed. For example, a table specifying the agricultural management of the fields where the quadrats were found might be added, from which questions about the impact of management practices on conservation value can be constructed. These types of question can be illustrated with reference to Figure 2.5.

A final point to note about these examples is that not all the data need be collected directly by the same individual or in the same manner or for similar purposes. In fact it is very likely that you will be relying at least in part on other people's data. The management data may have been collected to maximise profits, the species data as part of a study on the effect of climate on biodiversity and the data on conservation status to satisfy a legislative need, but *you* are trying to relate management to conservation value.

Time

Many GIS are rather weak in how they treat attributes that vary over time. Changes over time are almost always represented in a GIS as a series of discrete 'instances' or events separated by regular or irregular intervals during which the nature of change is undefined. Treating time and change as discontinuous is conceptually limiting. In many cases careful selection of the time interval can minimise the problem. Attributes that change size and shape cyclically or irregularly, such as seasonal lakes, the shoreline or the home range of an animal, are less successfully portrayed using a discrete schema.

Box 2.3   Types of search and example questions for relational databases

Class of search	Example question
One-to-one	Does the quadrat with the identifier 'b3' contain 'dandelions'?
One-to-many	List all the species found in quadrat 'b3'
Many-to-one	Find all the quadrats that contain 'dandelions'
Many-to-many	Find all the species found in at least two quadrats

The types of relational searches can be illustrated with reference to Figure 2.4.

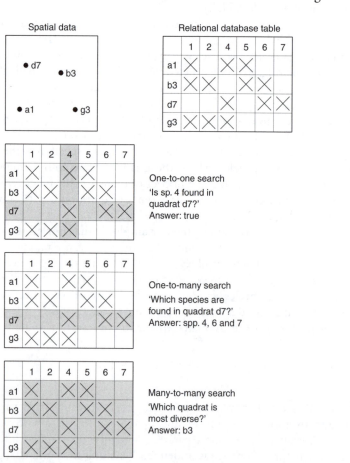

Figure 2.4   Illustration of searches of a relational database.

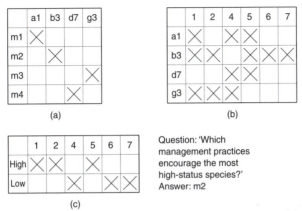

Question: 'Which management practices encourage the most high-status species?'
Answer: m2

Figure 2.5 Illustration of a more complex search of a relational database: (a) table management by site, (b) table site by species and (c) table conservation status by species.

Data models

GIS store information as a series of discrete units in an underlying database or 'data model'. What a computer stores is always a 'representation', 'model' or approximation of the real world, based on input data. Like all models, 'data models' must contain enough detail to be useful, but be simple enough to be useable and easily manipulated.

The 'data model' is the key to understanding how GIS software works. Within GIS software the most common data model is the 'field' model, which is concerned primarily with *geographic space*. 'Object' models are gaining in popularity; these are concerned primarily with *geographic place*. Field models are conceptually simpler and the examples in this book assume a field model. Many 'object-oriented' GIS are in fact used in an identical manner to a field model and we have yet to come across a problem that could only be tackled by an object-based model.

'Field' data models (geographic space)

The basic idea of a 'field' data model can be seen in Figure 2.6. In this simple example the real world is represented by three layers (or **coverages**) each one representing a particular 'theme'. These may also be referred to as 'thematic layers'. The first layer represents topography or elevation (in this case portrayed as **isolines** or **contours**), the second layer represents different types of vegetation and the third layer represents the territory boundaries for certain species.

Each thematic layer records information about the presence of attribute data. Each layer covers the entire area of the study (even if it contains no data for a particular location). Within a single layer the value of an attribute at a single point consists of a single value. Identical features in different parts of the

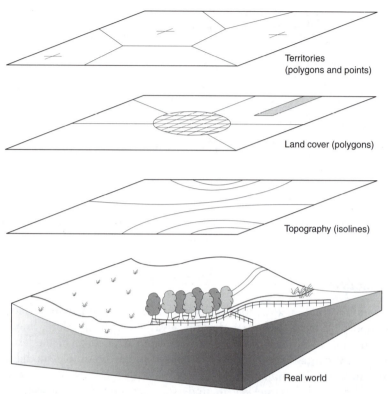

Territories
(polygons and points)

Land cover (polygons)

Topography (isolines)

Real world

Figure 2.6   Concept of using layers of data to represent the world.

same layer are considered to be identical. For example, on a layer representing woodland a patch in the north-west corner of the study site is considered to have exactly the same properties of 'woodiness' as a patch in the south-east corner. If the two patches need to be distinguished in their 'woodiness' (say by species richness) that needs to be made by an *explicit* combination with another layer.

Layers can be compared directly with one another because they share a common locational framework. In other words, every 'observation' included in the data model is tied to a specific location. This makes it possible to study and measure the positions of observations with respect to each other and therefore to search for possible spatial relationships and correlations between different sets of attribute data or thematic layers to answer some of the questions posed in Box 1.3 (using the techniques described in Chapter 5).

'Object-oriented' data models

Instead of looking at geographic space with a 'field' data model it is possible to consider geographic *place* and use an 'object' data model. Object-oriented

(OO) data models require a particular way of looking at the world. As a non-geographic example of an object-oriented data model: ecologists could form an 'object', a particular instance of this object is Jo. Another 'object' could be book publishers, of which Longman is a particular instance. One type of relationship between the 'ecologist object' and the 'publisher object' is a contract, which happens to exist between Jo and Longman. As you can see, each object can contain many instances and there is a multitude of relationships between objects. Each object may also contain a number of subsidiary objects (or classes or instances).

An OO conception of a landscape might contain an object 'woodland'; each patch of woodland in the study area would then be a unique instance of that object type. The concept of 'woodiness' may be expressed through defining attributes (such as 'trees') or by defining subsidiary objects, such as trees, which in turn may have attributes (or further objects) of roots, trunk, branches, leaves, flowers and so on.

Although there are conceptual advantages of an OO approach, in practice we find less to recommend it. The problem we have with the approach is 'seeing' the boundaries between objects; for example, is the object the wood or the tree? More formally Goodchild (1990) identified the following problems: many geographic constructs are implicitly uncertain, spatial objects are often the products of interpretation or generalisation and it is difficult to form a world of rigidly bounded objects. For an enthusiasts view of OO an interested reader is referred to Egenhofer and Frank (1987), Worboys *et al.* (1990) or Worboys (1994). Several GIS that employ an object-oriented approach exist, and they may be gaining in popularity; two fairly widely used examples are SmallWorld and ArcView.

Interpreting input data

It is important to realise that the information recorded in each 'thematic layer' of a field data model will have been manipulated to varying degrees. For example, referring back to Figure 2.6, data describing elevation may have been derived by **interpolation** from ground survey points, from overlapping (**stereoscopic**) pairs of air photographs or from stereoscopic pairs of satellite images. Elevation data can be considered to be (reasonably) objective because they will change little over time, are clearly defined and easily amenable to checking. On the other hand, derivation of the vegetation 'types' or 'classes' shown in the second layer involves a greater degree of interpretation of original data, for example, using 'expert judgement' or statistical analysis, and is much harder to verify. What the GIS stores in the thematic layer called 'vegetation types' may therefore already include errors due to misinterpretation of the 'raw' data; for example, when does scrub becomes woodland? Or when does woodland becomes a forest? The final layer in our data model depicts territorial boundaries for certain species and is even more uncertain. There may be inadequate field data to define boundaries with any degree of confidence

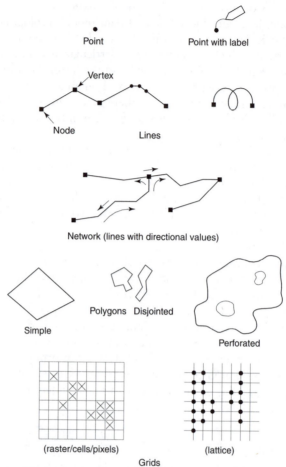

Figure 2.7   Fundamental structures of a field data model.

and the boundaries may also change frequently, such that this layer can only be regarded as a snapshot in time. Interpretation of 'raw' data is considered in more depth in Chapter 4.

Data storage in field models

Within a 'field' data model, data are stored using three fundamental structures: points, lines and grids; lines, which have **topology**, can be used to construct networks and **polygons**. Figure 2.7 shows these basic structures.

Points

A single co-ordinate pair used to represent features at a discrete location is termed a point. A distinction needs to be drawn between:

- Entities that are effectively reduced to a 'point' because of the scale of the 'map' (for example a town might appear as a distinct shape on a 1 : 50 000 map, but only as a point on a 1 : 1 000 000 scale map).
- Measurements relating to a discrete location (for example elevation and rainfall are both continuous variables that are regularly represented by values at particular points).
- Entities that have no real units of measurement or areal extent, such as the centre of a territory or the location where an interaction occurred. In systems where connected lines are considered to enclose map 'polygons' these can contain a 'seed point' or 'label point' that may be used to connect the polygon with attribute data. A seed point may be located anywhere within the polygon and does not represent a geographic feature.

In a GIS points at the end of lines or at junctions are termed **nodes**; points where a line changes direction are **vertices**.

Lines

Lines consist of joined co-ordinate pairs. They might be used to represent:

- linear features like streams or hedgerows,
- physical or implied boundaries, and
- to link points that have the same value such as *contours* or other types of *isoline*.

Lines in a GIS are sometimes referred to as **vectors** because they have a direction (although this is not a strict mathematical definition). Provided that information on *connectivity* is available, lines can be used to represent *networks*. Connected lines may enclose polygons, or areas that are homogeneous; in other words, everything within the area has a common characteristic or attribute. In ecology, it is common for areas or polygons to be organised hierarchically. For example, a stand of birch trees (homogeneous with respect to tree species) will be one area that can be found within a larger woodland (homogeneous with respect to land use, but heterogeneous with respect to tree species).

Grids

A raster grid is a regular matrix or two-dimensional array of cells. Cells are normally square but are occasionally rectangular. The location of each cell is implied from the extent of the grid and the cell size rather than being explicitly stored. Each cell stores the attribute data for that particular location. Attribute data can be accessed directly by geographic location or by row/column notation (the latter being very convenient for some forms of spatial modelling). A raster grid used to contain data from remote sensing scanners is sometimes referred to as an **image**, each cell being referred to as a **pixel** (or picture element). A raster grid can be interpreted in two ways: as a surface, or as a grouping of discrete categories. These two possible interpretations are

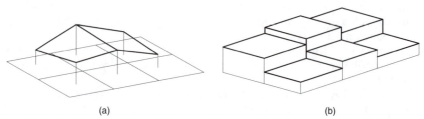

Figure 2.8   A grid may be interpreted as a continuous surface (a) or as a categorical (stepped) surface (b).

illustrated in Figure 2.8. A **quadtree** is a special form of raster grid that uses a variable cell size, with a finer division of cells where there is more detail and larger cells where there is less detail. The name quadtree comes from the process of dividing selected cells into four quarters.

Representing ecological data

There is a variety of ways of representing ecological data in a GIS and no universal 'right' way. Ecological observations in a study area may be represented in a variety of ways:

- a collection of irregular point samples, arising from randomly placed quadrats, the position of trees in a naturally regenerating wood or observation of an individual animal. (Note that by irregular we do imply that there is no pattern, only that it is not possible to exploit any geometric property of the pattern in the storage of the information.),
- a collection of regular point samples, arising from systematically placed quadrats or transects,
- lines delineating boundaries, arising from landscape surveys, or interpolated from point observations,
- a regular grid, arising from remote sensing, within quadrat measurements or **scanning** existing maps,
- polygons (lines drawn around the edge of a homogeneous patch). 'Homogeneous' polygons may represent a quantitative or qualitative interpolation from observation points. For example, you might stand in the middle of the study area and say "that area looks uniform, let's draw a line around it", or you might make a series of measurements that lend themselves to mathematical interpolation.

Two additional methods of storage of continuous data that always depend on *interpolation* might be used:

- isolines (lines joining equal values), in a process analogous to drawing contours you might draw isolines through point data to represent say biomass density or number of species,
- triangular irregular networks, where the resultant surface might represent the same sort of phenomena as isolines.

(Interpolation is discussed in more detail in Chapter 5.)

A considerable literature exists concerning the relative advantages of vector and raster GIS. Such arguments are increasingly sterile, as modern GIS software can

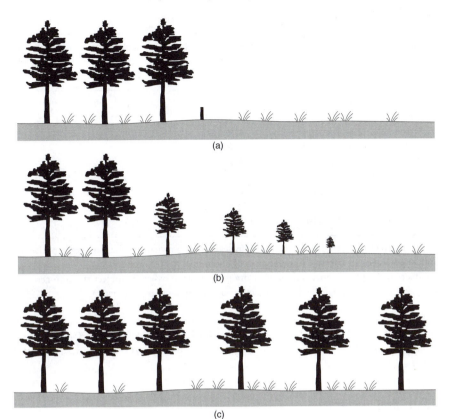

Figure 2.9 Different types of boundary: (a) abrupt, (b) gradual, and (c) statistical.

generally cope with both, though it still tends to emphasise one type. The choice of which representation system to use may be determined by three factors:

- how the data were collected (complete census or some form of sampling),
- your conceptual model of how the attribute might be changing (is it continuous, discrete, smoothly or abruptly changing, stable or chaotic?), and
- the ease with which subsequent analysis or modelling may be carried out.

It often makes sense to store data in a manner that reflects the way in which they were collected. For example, store:

- random quadrats as irregular points,
- results of vegetation or habitat survey as a set of polygons,
- remotely sensed data as a regular grid, and
- systematically placed soil samples as regular points.

Methods for representing data might also be influenced by the ways in which attributes can be expected to change. If you consider that an attribute is likely to show a gradual and even change over a study area, it might be preferable to represent it by using isolines. If, on the other hand, the attribute changes abruptly or in 'steps', you might prefer to use polygons. Figure 2.9 shows

Table 2.2   Scale, coverage and resolution

Scale	Area covered by an A4 sheet[a] (km)	Area that can be conveniently shown on an A4 sheet[a] (km)	Resolution – minimum size of object or separation of objects[b] (m)
1 : 500	0.148 × 0.105	0.10 × 0.075	0.01
1 : 1000	0.297 × 0.21	0.20 × 0.15	0.02
1 : 5000	1.485 × 1.05	1 × 0.75	1
1 : 10 000	2.970 × 2.10	2 × 1.5	2
1 : 25 000	7.425 × 5.25	5 × 3.75	5
1 : 50 000	14.8 × 10.5	10 × 7.5	10
1 : 100 000	29.7 × 21	20 × 15	20
1 : 250 000	74.25 × 52.5	50 × 37.5	50
1 : 1 000 000	297 × 210	200 × 150	200

[a] Leaving space for margins, titles, scale bar, legend, acknowledgements, copyright notice, key and so on leaves an effective drawing area of about 150 × 200 mm.
[b] This assumes professional cartographic draughtsmanship, where an object 0.2 mm across on the paper can be distinguished and drawn – but that takes years of practice.

some different types of boundary that might exist between a woodland and a grassland. These different types of boundary are best represented in different ways.

Scale and resolution

As we mentioned earlier scale is a term that often causes confusion; colloquially large and small scale usually refer to the size of the study site, while cartographically small and large scale refer to the ratio of the size of the map to the size of the site. Cartographically a 1 : 500 scale map is very large-scale, a 1 : 1 000 000 is a very small-scale map (because five hundredths is larger than one millionth). A small-scale map needs a small piece of paper to show a large piece of land; a large-scale map needs a large piece of paper to show the same area. The confusion arises with the phrase 'small-scale study': does it mean a study of a small area? Or does it mean a study of a large area at a relatively low level of detail? 'Small-scale' is not as ambiguous as 'bi-monthly', but it is always better to be explicit.

The resolution of a raster-based map is determined by and equal to the size of the grid cell. The resolution of a polygon, vector or a point map layer is primarily dependent on how the data are captured (the **precision** used to store the numbers within the computer is rarely of significance). Table 2.2 provides some examples of the amount of detail (resolution) and area covered by some common mapping scales.

The resolution shown in Table 2.2 provides an estimate of both how accurate you need to be to achieve a particular aim and also what is so small

Table 2.3   Correspondence between 'Imperial' and metric scales

Imperial	Scale	Closest metric scale
6 inch to one mile	1 : 10 560	1 : 10 000
1 inch to one mile	1 : 63 360	1 : 50 000
quarter inch to one mile	1 : 253 440	1 : 250 000

that it can be omitted. For example, in England the mean size of a Site of Special Scientific Interest (SSSI) is 75 ha. Depending on how compact a shape it has, then a scale of around 1 : 5000 would be convenient to show the whole area on an A4 sheet. If you are going to produce a map of an 'average' SSSI at that scale then strictly speaking you should determine the location of *all* points that are to be shown with an absolute **accuracy** of 1 m (which is not a trivial task), it also means that features smaller than 1 m across can only be shown by telling lies – exaggerating their true size or lumping several similar features together, or displacing the position of a feature if it is too close to another feature. Similarly a road atlas produced at say 4 miles to the inch (as the one in my car is) has a scale of approximately 1 : 250 000 and a corresponding resolution of 40 m – which means it cannot show truthfully any roads smaller than a dual carriageway (does this explain why Richard takes so many 'scenic detours'?).

Table 2.3 shows the approximate correspondence between 'metric' and 'Imperial' scales.

Aspects of data quality

Warning

Errors in data are *inescapable* and should therefore be considered as a fundamental dimension of GIS data. The objective should be to assess and quantify error so that valid conclusions can be drawn from the data available.

Never forget to consider the quality of input data. Much heartbreak, trouble and stress can be avoided by asking a few simple questions about any piece of data you want to use:

• Who collected the data?
• When did they collect it?
• Why did they collect it?
• How did they collect it?

Even respected research bodies have been known to issue data with no indication of their accuracy, precision or reliability, so there is absolutely no guarantee that you will be able to find answers to all these questions, but it is at least worth trying. Terminology commonly encountered in discussions of data quality are listed in Box 2.4.

Box 2.4   Common terminology used in discussion of data quality

accuracy – deviation between a measured value and the true value

attribute accuracy – accuracy of the attribute

completeness – whether the digital data cover all the relationships and objects which they are supposed to

currency – the degree to which data conforms to the time period they are supposed to

local or relative accuracy – accuracy between neighbouring points

logical consistency – whether the representation of objects (within the GIS) is correct; for example, if a river is represented as a network of lines do the correct lines join at the confluences

global or absolute accuracy – accuracy relative to an absolute frame of reference

positional accuracy – accuracy of the location of a point, line, etc.

precision – amount of 'detail' or number of significant digits in a measurement: a distance measured to the nearest millimetre is more precise than one measured to the nearest centimetre (but not necessarily more accurate)

resolution – smallest difference that can be distinguished

Accuracy and precision

It is important to maintain a clear distinction between accuracy and precision. The root of the word 'accuracy' is in the Latin verb 'to take care of'. Accuracy is a measure of the degree to which a value is free from error or how close it is to a representation of reality (the truth). The root of the word 'precision', on the other hand, is from the Latin verb 'to cut'. Precision defines the number of significant digits used to specify a value. The difference between 'accuracy' and 'precision' can be illustrated effectively by reference to global positioning systems (GPS). A GPS records the locations of the receiver by **trilateration** from a swarm of satellites. A hand-held receiver will typically report position very precisely to fractions of a metre. However, the recorded position will only be accurate to several tens of metres, because of deliberate satellite signal scrambling, so you can be very precise and wildly inaccurate at the same time. It is also important to maintain a distinction between **local** and **global accuracy**, or between absolute and relative accuracy. One of the nice things about GPS, for example, is that spatial errors are not auto-correlated: how ever close or far apart the points are, the uncertainty in their relative position is likely to be the same. The error in the distance estimated by GPS between London and New York will therefore be of the same magnitude as the error in estimating the distance between the Empire State Building and the Statue of Liberty. On the other hand, positions determined from 'ground' or 'field' surveys do tend to be spatially correlated: the further points are apart in reality, the greater the uncertainty about their relative positions is likely to be.

The term accuracy is also used in connection with how well a point or other feature is described, the **attribute accuracy**, which is (normally) how closely the attribute describes the actual phenomena being represented. The attribute may have too few classes (or the classes too poorly defined) to define the phenomena accurately. Poor attribute accuracy may also arise because the data are old or collected in an inconsistent manner (including the **logical consistency** of the assignments) or attributes are incomplete or the method of assigning the phenomenon to an attribute is poor. Attributes that are assigned during a well-defined time period are said to be *current*.

The attitude of land surveyors to error and uncertainty provides a robust and realistic view of spatial accuracy and precision. The first thing a land surveyor considers before collecting any data is how accurate the final result needs to be for its intended purpose. They do this by working back from the needs of the end user, the product they require, and the various presentational, analytical and data-manipulation processes needed, to the techniques to be used for data capture. To illustrate this process, suppose that you wanted a land surveyor to provide a map showing the annual increase in bracken on your nature reserve. A patch of bracken might be spreading by between say 0.25 and 0.5 m per year. To be certain of reliably detecting (and quantifying) change it will be necessary to 'fix' (determine) the position of the boundary to a finer level of accuracy than the potential change, in this case say 0.1 m (100 mm). Because you have asked a land surveyor to do the task, the 100 mm requirement also determines the mapping scale to be around 1 : 500. The requirement for determining position to 100 mm next determines the technique to use: GPS is nowhere near accurate enough, air photography might be possible but you would need photographs at a scale of about 1 : 3000 (which would not be cheap), so that leaves some sort of ground survey method. Using a theodolite and electromagnetic distance measuring (EDM) it is possible to measure position (from distance and angle) to within a few millimetres; however, it is necessary to allow for the theodolite not being exactly over the 'survey station' and the 'pogo stick' (the pole with the prism on it) not being exactly over the boundary then we might be able to determine the location of the boundary to within say 25 mm. The next thing to determine is the minimum set of 'survey stations' from which all the points that need to be surveyed can be seen, and at the same time it must be possible to link all the survey stations together by further measurement (probably through **traverse**). Because we have set the overall accuracy at 100 mm we must determine the relative position of the 'survey stations' so that the total accumulated error is less than 75 mm.

As a second example suppose you require to relate the position of beetle pitfall traps to land cover estimated from the Land Cover Map of Great Britain (LCMGB). The LCMGB has a resolution of 25 m. It is known that the process of fitting the satellite image to national co-ordinates may introduce an error of up to one cell. So the position of the pitfall needs to be known to better than half the cell size (12.5 m). This means that the use of a hand-held GPS is probably not going to be good enough. Differential GPS would be accurate enough. On the other hand all of Britain has been mapped by the Ordnance Survey (OS) at 1 : 10 000 scale, so identifiable points on an OS map should be known to within a couple of metres. This means it should be sufficient to measure in from identifiable points with a chain or a tape as you can afford to have an error of up to 5 m. Five metres sounds like a large margin but in upland Britain, points that can be unambiguous are often so far apart that a simple tape or chain is going to involve a lot of work and require very careful use (keeping the line straight, adjusting for slope and so on) to reach the required accuracy.

Sources of error

There are many potential sources of error. Several classifications of error are used but the system described by Burrough (1986) is robust and widely used. This divides sources of error into three classes:

- obvious sources of error,
- errors due to natural variation or method of measurement, and
- errors that arise through processing.

Although it can be argued that conceptual errors are a 'processing' error we would like to suggest that conceptual errors merit a distinct class of their own.

Conceptual errors

The way a set of data is conceptualised can have a significant effect on the reliability of any conclusions that can be drawn from it. Conceptual errors are, of course, very difficult to quantify or remedy. The way you view the world can be so ingrained that often you do not realise the influence it has. Whereas one ecologist might view the landscape as a series of distinct homogeneous patches, another might conceptualise it as a series of gradually changing ecotones. Similarly the changes that occur in a mosaic of patches may be conceptualised as gradual and continuous with fixed rates of transition, or as periods of equilibrium interspersed with occasional 'catastrophic' events. It is also possible for conceptual model development to be driven or 'led' by available or easily accessed information.

Obvious sources of errors

'Obvious' sources of error include the age of the data, their consistency of coverage, their scale, the density of input observation, their relevance, their format and their accessibility and cost. The first four sources of 'obvious' error are often inter-linked: for example, the density of observations can be expected to be related to the scale and possibly the age of the data.

Age, consistency, scale and density of observations
There are a few lucky people who can capture all the data they need when they need it. The rest of us often have to make do with 'old' data. As well as

the date of the data there can also be problems with the season in which they were collected. Although old data can sometimes be very interesting (and valuable) records of ecological phenomena it is important to remember that techniques, definitions and systems can all change; also the cost of capturing information from field sheets can be enormous. When a study covers a large area it is not uncommon for different sections to be inconsistent. Inconsistency most often becomes manifest where map sheets join. As a general rule, consistency is better achieved by generalising the complex than 'blowing up' a small-scale map to achieve coverage. Estimating the optimal sampling density for a particular phenomenon is not a trivial exercise; however, knowledge of the sampling density can provide a very rough guide to data quality.

Relevance

It is not uncommon to have to make use of surrogate data as the data needed are too difficult or expensive to collect. The science of remote sensing is almost entirely based on the principle of surrogates: measures of vegetation cover, mass and density, for example, are all derived from a bit of reflected sunlight! There are several ways to establish the relationship between the surrogate data and the required data, such as, regression analysis, co-spectral analysis, co-kriging and so on, which will be discussed in Chapter 5.

If information is being captured from an existing map it is necessary to appreciate some of the cartographic processes that the data may already have gone through. Information may have been selected, simplified, displaced, smoothed, or exaggerated. These issues will be discussed in more detail in Chapter 4. The possibility of encountering a 'fake' object that has been inserted to catch copyright thieves is rather remote but not necessarily negligible.

Format

The term 'format' can be applied to three aspects of the data: how it is physically stored (tape, disk, ASCII, binary), how measurements are represented (points, lines, grids, etc.) and how the measurements are formed (scale, projection, classification). Errors arising from the first use of the term 'format' usually result in complete garbage being generated, or complete failure to extract any data, both of which are easily identified. Errors due to the way measurements are represented are perhaps more conveniently thought of as conceptual errors.

It is the third aspect of format (and particularly the projection and classification used) that is most likely to generate errors. Provided that you know exactly what the projection is and what all the parameters are, **conversion** between different projections can be easily accomplished by applying the appropriate mathematical transformations (which are included in many GIS). Do not be tempted to apply an approximate transformation using some sort of 'rubber sheet' method; newsgroups concerned with GIS contain a fairly constant series of requests for information about the projections used in various regions at various points in time and someone will usually reply within a day or so. Converting between different classification systems can be a nightmare.

In our opinion the different vegetation classifications are really bad, but there are those who argue that soils are much worse. Various data dictionaries comparing existing classifications have been produced, e.g. Wyatt *et al.* (1997a, b), who investigated 15 classifications from the USA and Europe or Cox and Parr (1994), who compared ten(!) classifications used in Scotland. This problem of inter-comparability has prompted the Food and Agriculture Organisation (FAO, 1997), for example, to provide a framework for the construction of country- and region-specific vegetation classifications by encouraging the use of the same vocabulary and conceptual framework.

Accessibility and cost
In our opinion these are not strictly sources of error, but if you cannot get hold of the data or cannot afford it then analysis is not going to be as good as it might have been.

Measurement error

Measurement errors include positional, content, classification and bias. Some phenomena are naturally variable in time and space to the extent that an exact representation is not possible.

The two primary measurement errors are related to location and content.

Positional errors
Where features are man-made, solid, physical, easily identifiable entities, such as buildings or roads, not only can their location be defined accurately and precisely but they can also be checked easily. Therefore the accuracy of topographic maps typically relates to these 'surveyor friendly' features. Features and boundaries likely to be of interest to an ecologist are much more difficult to define (in space and time) and so their locations are more likely to be subject to uncertainty.

Content errors
These might be such things as misidentifying a species or under- or over-estimating the dominance of a species.

Accumulation of measurement errors
Having decided on a method of data collection it is necessary to organise the data collection to guard against the many sources of error that can creep in. Consider estimating the total error of adding two quantities A and B together. If A and B are completely independent of each other and if the error (uncertainty) associated with each of them follows a Gaussian (normal) distribution then the cumulative error is the square root of the sum of the uncertainty. For independent observations with 'well-behaved' errors there is an equal probability that an observation will be greater than or less than the true value. However, if the error structure is not symmetrical (is biased in some way) and

the observations are not independent then the cumulative error may accumulate much more rapidly.

Processing errors

Errors in processing include numerical and hardware problems, logical and topological errors and classification and generalisation problems.

Numerical and hardware problems

With early computers a lot of thought had to be given to the precision with which a computer stored numbers and the numerical stability of any method of calculation. The classic examples of numerical instability in computers involve repeatedly squaring a number very close to unity (leading to rounding errors) or adding and subtracting sequences of numbers that are very different in magnitude (leading to violation of the rule of commutativity). Some people like to work in double precision at all times and hope for the best; we prefer to use integers wherever possible and to carry out a few trial calculations by hand first to make sure that we do not end up by dividing by a number very close to zero or using the difference between two very large but very similar numbers. It is also worthwhile to be very cautious about testing for equality with real numbers; if necessary be explicit about the acceptable rounding error or uncertainty.

Errors in the overlay process

Overlaying information is one of the classic GIS operations; one of the things that is often neglected is the way that errors can increase. For example, suppose we wish to relate soils type to vegetation type. In the USA the definition of a soil map unit states that inclusions are allowed to cover up to 25% of the map unit provided they would not substantially affect soil management (Soil Survey Staff, 1983), so with uncertainty in the boundary between soil types and the actual classification of the soil the accuracy is not likely to be better than 60–70%. Bolstad and Smith (1995) report an overall accuracy of 67.2% in distinguishing between conifer, hardwood and mixed forests using remote sensing. Therefore our overlay might have an accuracy as *high* as $0.672 \times 0.7 = 0.47$, i.e. less than half! This result is very problematic. It shows that when using an overlay process it is not very difficult to end up with a composite map that on the face of it has virtually no accuracy at all. It should be noted that such a result is counter-intuitive. We normally assume that the more *evidence* we have the more *certain* we should be. There are methods for dealing with data as evidence, but such approaches are not very widely used.

Modifiable area unit problem

One of the thorny (and much discussed) problems with spatial data is the modifiable area unit problem. An interested reader is referred to Openshaw (1984) for a detailed discussion, but in essence, *any* pattern observed across a zone may be as much a function of the zone boundaries as the underlying spatial distribution of the attribute. Attempts to re-establish the underlying distribution are investigated by Robinson and Zubrow (1997).

Generalisation

Generalisation is tied up with how we represent something. When you go out using your quadrat to describe the vegetation of an area you are saying that you can represent the important characteristics of the whole field by measuring just a tiny percentage of the total. Similarly as you digitise a river network the choice of the points to represent the path is a form of generalisation.

Generalisation is a one-way process and hence processes that generalise data should be performed as infrequently as possible. Cartographic literature is replete with ideas for 'optimum' generalisations that destroy the minimum amount of information. However, in the case of map production the aesthetic qualities of generalisation cannot be discounted.

Another form of generalisation is filtering. Smoothing filters lose information. Vieux (1995) discusses the loss of information in terms of changes in entropy, using information theory, which you may recognise from Shannon's measures of diversity. In the case of discrete variables the entropy equals:

$$I = -\Sigma P_i \ln(P_i)$$

where P_i is the probability of a variate occurring in a discrete interval and the summation is over the total number of 'bins'.

Vieux (1995) also discusses the implications for this loss of information with hydrological modelling.

Landscape metrics and misclassification errors

There has been considerable interest in landscape ecology in measuring the size, shape and connectivities of patches in the landscape. A number of studies exist that try and link the population of various species (forest birds seem particularly popular) with various metrics of pattern. Any classification is subject to uncertainty: what effect does this have on measures of landscape pattern? Remotely sensed (RS) data from satellites form an increasingly important source of data for classifying landscapes. Correspondence between classified data and ground surveys is obviously dependent on a large number of factors but is rarely better than 70 or 80%. Experiments were carried out by Wickham *et al.* (1997) to investigate how misclassification errors will affect measurements of landscape. Three landscape metrics were examined in the study: average patch compaction (APC), contagion (C) and fractal dimension (F); they were chosen after factor analysis showed the measures to be close to orthogonal. Simulated errors were generated for a variety of landscape types based on Landsat Thematic Mapper (TM) imagery. Misclassification always lowers the value of APC and C, and always increases the value of F. With a misclassification rate of 12% a difference in the metrics of 17% was needed to be confident that real differences exist; such results would appear to be general for 'raster' landscapes.

Classification errors

Classification is a process very frequently undertaken in ecology. In the context of GIS and ecology, classification may arise in a number of ways: visual (manual) classification (or interpretation) of air photographs, supervised and unsupervised classification of remotely sensed data, manual classification or interpretation of sample data and automatic classification of sample data.

The accuracy of photo-interpretation obviously depends on what is being mapped and the resolution or scale of the imagery; as an example Bolstad and Smith (1995) report an overall accuracy of 67.2% in distinguishing between conifer, hardwood and mixed forests.

The accuracy of supervised and unsupervised classifications of remotely sensed data obviously varies with the number of classes defined and the resolution of the imagery and whether more than one image is used. Pathirana (1990) reports an overall accuracy of 77% in classifying Landsat imagery of land in the suburban zone into seven classes (water, buildings, asphalt, open land, bare ground, woodland and wetlands). Highest agreement with ground truth was for the water and wetlands classes (which might be expected to tend to occur in large homogeneous patches), lowest agreement was for buildings and open land (which might be expected to occur in small but heterogeneous patches). Conversely Edwards *et al.* (1998), in discussing the problems of assessing the accuracy of large areas of classified remotely sensed data (in their case the whole of Utah), found the highest agreement for man-modified surfaces (90.6%), and the lowest agreement for barren and aquatic surfaces (50.4 and 52.4%, respectively), with forests, woodlands, shrublands and herbaceous surfaces having agreements between 73.3 and 77.2%.

Soil surveys provide an interesting example of manual classification: although some data are collected from soil pits and auger holes, much of a soil map is an intuitive interpolation by the soil surveyor from the topography and land cover. In addition to errors introduced by the interpolation, each class may contain a number of subclasses that may have very different properties. In the USA the definition of a soil map unit states that inclusions are allowed to cover up to 25% of the map unit provided they would not substantially affect soil management (Soil Survey Staff, 1983).

Enough has been written about classification using various multivariate statistics of non-spatial sample or attribute data using all sorts of software packages to need to write anything here: Krzanowski (1988) and Reyment and Joreskog (1993) provide solid introductions and applications.

Spatial processes in ecology

There are several ways in which ecological processes occurring over space can be conceptualised. Common conceptual models that can conveniently be described in a GIS are point processes, continuous surfaces, network and zone (or area) processes. Spatial analysis will be discussed in more detail in later chapters but it might be useful to give some examples of ecological spatial processes here.

Point processes in ecology

Possible processes are:

- Distribution of individual plants across a field or trees in a wood. Such points might be clustered (limited dispersal capability away from one or more 'parents'), or they might be evenly distributed (competition for resources preventing establishment close to existing points), or they might be randomly distributed.
- Randomly placed quadrats in a botanic survey.
- Observations of an individual animal (even radio tagging is likely to be processed as discrete points although successive points will not be independent).

Continuous 'surfaces' in ecology

Possible processes are:

- An ecotone between two different habitats (change will be gradual but not necessarily uniform as in a salt marsh between terrestrial and marine ecosystems).
- A cost surface representing accumulated resistance to movement (the surface may include discontinuities or barriers to change).
- Biomass or photosynthetic activity across an area (may be the product of a mixture of uniform and non-uniform environmental factors, e.g. climate, soils, etc.).

Networks in ecology

Possible processes are:

- A river network can be represented as a network within which movement by organisms may be more or less difficult in different directions.
- Flow of genetic material between discrete populations (works on genetic rather than geographic space).
- Optimum routes for the movement of animals across a landscape between feeding and resting areas (could be derived from an accumulative cost surface).

Zones and areas in ecology

Typical processes are:

- Typical vegetation map, which consists of 'homogeneous' polygons.
- Patches within a forest that were burnt at different time periods or that were subject to outbreaks of pests at different times.
- Unique combinations of a number of environmental variables, such as the intersection of soil types, topography and vegetation cover.

Review

In this chapter we introduce what we mean by spatial data and the difference between spatial data and attribute data. How data are represented and stored within a GIS influence how a problem or environment is conceptualised and what sort of questions can be answered. As well as the conceptual problems of how data are represented, error and uncertainty are fundamental aspects of *all* data; what is important is that some appreciation of the sources and causes of error is maintained.

Further reading

As well as P.A. Burrough *Principles of Geographic Information Systems for Land Resource Assessment*, Monographs on Soil Resources, Survey No. 12 (Oxford Scientific, 1986), M.F. Goodchild and S. Gopal *Accuracy of Spatial Databases* (Taylor & Francis, London, 1989) provides a useful compendium of work on accuracy. A collection of papers on multivariate statistical techniques specifically for use in ecology can be found in R. Jongman, C. ter Braak and O. van Tongeren *Data Analysis in Community and Landscape Ecology* (Pudoc, Wageningen, 1987), or K.V. Mardia, J.T. Kent and J.M. Bibby *Multivariate Analysis* (Academic Press, London, 1979). Several Internet Web sites demonstrate different types of map projections, including **http://ww.aquarius.geowar.de/omc** and **http://everest.lunter.cuny.edu/mp**. Information in a more 'traditional' medium can be found in L.M. Bugayevskiy and J.P. Synder *Map Projections: A Reference Manual* (Taylor & Francis, New York, 1995) or J.P. Snyder *Map Projections A Working Manual* (USGS, Professional Paper 1395, Washington DC, US Government Printing Office, 1987). D. Dorling and D. Fairburn *Mapping: Ways of Representing the World* (Addison Wesley Longman, Harlow, 1997) provides an interesting (and up-to-date) discussion on how the world is portrayed and some of the personalities who have been influential in how we all perceive the world. Technical details of how GPS work can be found on the US Coast Guard Web site **http://www.navcen.uscg.mil/gps** – for the very keen it also supplies a bibliography of more than 1000 entries (although some are rather elderly!).

Chapter 3

Getting your project started

Defining what you are trying to do

GIS are often 'sold' as a tool that will help integrate and co-ordinate all sorts of information – and so they can – but integration and co-ordination of information is never a trivial activity. Successful integration and co-ordination can occur only if *everyone* involved in the project believes that they must actively foster co-operation and integration from the very *beginning*. Even if the project staff only consists of one person – you – you cannot leave planning the GIS activity to the end and just hope that it will all come together (this is our bitter experience). As Chambers (1983) points out in the general context of planning projects:

> Both integration and co-ordination have high costs. Both involve choices by default – choices not to use funds, administrative capability and staff time, in other ways. . . . There may be a law that the chances of a report . . . being implemented vary inversely with the frequency with which the words integration and co-ordination are used. For they evade the hard detailed choices of who should do what, when and how, which are needed to make things happen.

It would be a mistake to try and start to use GIS without doing some initial planning. In a study of 500 large GIS projects (involving more than 25 person years of effort), 15% were aborted or not used at all and a further 25% were judged to have failed (Huxhold and Levinsohn, 1995). Technical problems were rarely important in the failure; the most common reason for computer projects to fail was a lack of clear objectives as to what was to be done. There appear to be no comparable studies on how and why small GIS projects fail, but there are no reasons to suppose they have higher success rates than large projects.

The two questions that you must ask before you start are:

- What am I trying to do?
- What resources do I have?

Nine principles for successful implementation of a GIS project were proposed during a joint meeting between the International Association of Assessing Officials (IAAO) and the Urban and Regional Information Systems Association (URISA) (reproduced opposite from Huxhold and Levinsohn, 1995). Note that these principles were devised to describe implementation of GIS in large government organisations; some of the principles (especially the last one) are unlikely to be relevant to a lone ecologist or a small group of ecologists. The value of the principles lies in the fact that they were derived from an empirical study of 'GIS practice' and not from a theoretical or 'GIS science' perspective. Some of the principles give (we hope) a glimpse of the blindingly obvious, in particular principles 1, 2, 3, 4 and 6, which should be relevant to every project. Italics have been used for sections we wish to emphasise.

Nine principles of the IAAO and URISA (Huxhold and Levinsohn, 1995)

Principle 1: A GIS is a data-driven data-based information system.
Making maps is only one capability of GIS technology. GIS is based upon database management concepts that allow flexible access to data, interrogation of data from different sources and manipulation and analysis of data in ways that are not possible with other techniques or technologies.

Principle 2: GIS data and maps *must* be maintained.
Because the GIS is designed for use in the operating environment of the organisation for specific service delivery, management and policy-related activities *the data and ultimately the system will not be used if they are out of date or if they are inaccurate.*

Principle 3: A GIS is most useful when geographic references are registered on a consistent, continuous co-ordinate system.
A GIS is not merely a collection of computerised map sheets. Its use demands that an entire geographic area be accessible in order that the spatial relationships between features in different parts of an area can be identified and displayed. This requires locating each map sheet in its proper place within the geographic area. It is accomplished by setting co-ordinates for each map sheet that come from a consistent, continuous co-ordinate system.

Principle 4: A GIS has topology.
Because a computer cannot see a map as a human can, additional definitions of the relationships between points, lines and areas must be established. This topology allows a GIS to perform certain spatial analysis functions, including (but not limited to) network analysis and optimal path determination, polygon overlay, geo-coding and area calculations and shading.

Principle 5: A GIS has many uses and should be shared by many different functions.
Because the value of information increases the more it is shared and used by others who need it, and because GIS requires significant resources to develop, a multi-use, shared system can prevent duplication of common data and the effort to maintain them and can also reduce the cost of the system to any single user.

Principle 6: A GIS contains hardware and software that are constantly undergoing change, which improves its functionality over time.

A delay in acquiring GIS hardware and software in anticipation of future price reductions or technological breakthroughs is not prudent, because existing technology is fully adequate to develop GIS applications. The benefits of the system cannot be realised until the database is built and implemented on the system, so a delay in implementation only creates a delay in realising its benefits. Future improvements to technology will enhance the use of the data – not restrict it.

Principle 7: A GIS grows incrementally in terms of technology, cost and administrative support needed. Therefore, a long-term commitment is needed to assure success.

The large amount of time required to build the databases and the large number of potential users and applications prevent a GIS from becoming fully functional within a short time frame. *Given limited resources and an ambitious plan for GIS implementation, priorities must be established* and commitments maintained over a multi-year time frame.

Principle 8: A GIS causes changes in procedures, operations and institutional arrangements among all users.

The common databases accessed by many different users eliminates the compartmentalisation of data storage and individualisation of data-coding schemes. This will result in changes in responsibilities, procedures, security measures, standards and even organisational structures and laws in order for GIS to function for the benefit of all.

Principle 9: A cadre of trained, educated, motivated and dedicated people is crucial for a successful GIS program.

Without exception, organisations that have successful systems have been able to assemble, and retain for a long time, the appropriate level of staff with technical and communication skills who have, as well, a shared vision of the potential for the technology. Technical problems can be resolved with money and time, but staff without motivation, dedication, creativity and a willingness to accept new ideas are likely to scuttle the project. Most successful systems have a 'champion' – a high-level official who is willing to push the project forward and motivate and educate those whose support is needed. However, increasingly, successful organisation-wide GIS implementation is being led by a dedicated team, rather than a single champion.

Setting priorities

The first thing you need to do is set down your 'wish list' for your GIS and then set some priorities. It is important that you break down the process of putting the GIS together into a series of achievable goals. For each goal you need to set priorities before you start to implement any part of your system. Priorities imposed from outside are always a burden, but the ones you set yourself are always helpful. We believe in setting lots of little 'milestones' so that we have at least the illusion of making progress (for example, with this book the target was one paragraph per day every day – it was not a lot of fun but at least the illusion of progress was maintained!).

Of course there are many potential levels of priority. We would suggest that three are enough:

- Highest priority:
  —tasks that cannot be done without the GIS;
  —tasks that provide a 'big bang for the buck' (a lot of benefit for the minimum amount of input) – purists might disagree with this criterion, but especially early on in a project there is a great deal to be gained from performing a task that makes other go 'Wow!' and makes you realise that you *are* making progress; and
  —tasks that need to be done regularly but are very time-consuming and laborious to do manually.
- Medium priority:
  —tasks that are done regularly but infrequently,
  —tasks that can form building blocks for more complex activities later on.
- Low priority:
  —tasks that are needed irregularly and very infrequently,
  —tasks that are dependent on other things being implemented first,
  —tasks that are difficult to automate or that can be achieved relatively easily manually,
  —tasks that require a lot of effort for little obvious benefit.

By defining what your applications are you should be able to determine not only what software is appropriate and what hardware you need, but also what data you need.

Determining your software requirements

The problem with this section and the one following is that development of software and hardware is so rapid that anything we write could be out of date before you get to read it.

In Chapter 5 we make reference to three widely used GIS but you should be aware of just how many packages are available. During 1991 a census of software reached a total of 351 products (GIS World Inc., 1991). It would be difficult to impossible to give a comprehensive or even a reasonably complete overview. Because you are going to invest a considerable amount of your time in learning to use your GIS it is important that you make a considered choice. GIS systems can cost from next to nothing (for example GRASS, developed by the US Corporation of Engineers, can be obtained for nothing by downloading all the *source code* from the Internet and compiling it!), to tens of thousands of pounds for a commercial licence for some products. In our *opinion* correlation between price and performance is not strong for GIS software. Much GIS software is offered with very high educational discount as each vendor struggles to 'catch them young'.

In 1997 the journal *Mapping Awareness* published a software review covering 74 commercial organisations producing or retailing 138 different products; similar reviews are regularly produced by various journals. Anyone considering purchasing a GIS should try and obtain one or more recent surveys. Because of the speed with which such information ages there is little point in going through any of the reviews in great detail, but we will indicate the range of possibilities with regards to user base, cost and capability.

The number of licences associated with an individual product ranged from just 12 to 200 000; in 1997 four commercial organisations claimed more than 100 000 licences worldwide. Although it is difficult to form much of an impression of who is using GIS it can be noted that in the *Mapping Awareness* survey 'environment/natural resources' merits a specific qualitative assessment (along with utilities/telecoms/cable, local government, central government, business, transport and emergency services) but that education is lumped in with 'other'. The oldest product reviewed was launched in 1975 and nearly 30% (36 out of 127 with a specified launch) were pre-1990; however, nearly 40% (47 out of 127) of the products were launched in 1995 and 1996. Many products work on several computer operating systems, although the widest choice is for PC-based systems followed by UNIX and Macintoshes.

Just over one-third of the products (57) were not priced, but instead specified as price on application (POA). Of the companies that would quote a price, values ranged from free to US$20 000. Although many companies specified a range of prices there were 20 products for under US$1000, a further 39 products for under US$5000 but 24 products cost more than US$5000.

Given that these products are being reviewed in a mapping journal it is a bit surprising to discover that fewer than 59% (81) of them allow you to transform data in different co-ordinate systems and only 28% (38) offer raster–vector conversion capability. Even more surprising is that only 42% (58) include '. . . cartographic design features'. But to be more optimistic 81% (112) at least allow raster and vector data to be displayed at the same time, and just over half (72) have at least some analysis capability with both raster and vector data. Note that the newest software (launched in 1995 or later) is not significantly more likely to have dual capabilities than older software.

Given the plethora of choice, what are you going to do? Box 3.1 provides some suggested questions to ask before obtaining any GIS software.

The first and probably most important way to narrow the choice is by considering what existing expertise is available to you. It is possible to learn a GIS from reading the manuals and we know several people who have done so, but it is hard work, and after all you want to *do* ecology not learn software. The simplest GIS contain dozens of commands, the largest several hundreds. It is not going to be trivial to become a 'knowledgeable user'; if someone can give you some help on a particular piece of software it is a very strong incentive for starting there.

As the distinction between raster- and vector-based GIS systems is becoming increasingly blurred, there are still plenty of packages that are stronger with one sort of data than another. One of the most widely used packages at the moment can deal with raster data only as a pretty backdrop behind vector data. Other packages use vector data primarily to make the image look more 'map-like' but allow little analysis to be carried out with it. Given the way

---

Box 3.1 Suggested questions to ask before obtaining any GIS software

1. Do I know anyone who has *used* this software? If yes:
   Do they think it was a good product?
   Will they be willing and able to help me learn how to use it?
2. Is the software used by any potential collaborators?
3. Has the software been used for a similar project to the one I am proposing to carry out?
4. Can it handle *all* the types of data I might want to use: raster, vector, points, labels, text?
5. Can it handle the *volume* of data I am likely to generate?
6. Do I have to take it on trust that the commands do what they say they will, or will I have access to the algorithms used?
7. Will it run on hardware I am familiar with, or will I have to obtain new hardware, or learn a new operating system?
8. How easy is it to import data from other software packages (including databases, spreadsheets, statistical packages, graphical packages)?
9. Is it primarily a mapping (graphical) package or does it have a serious amount of spatial analysis functionality?
10. Can I afford it?
11. Is it likely that the software company will still be around to support, upgrade or 'bug-fix' the product in a few years time?
12. Does it allow batch files, shell files, macros or other means to make my life easy?
13. Does it do more than I think I am ever going to need?

---

projects tend to develop it is risky to assume that you would never need to use data in a particular format, so go for a system that is flexible.

Data storage is really a hardware problem and as storage becomes increasingly cheap it is less of an issue than it was a few years ago. However, the subtlety with which software deals with large data sets does vary a lot. The last thing you want to have to do is think about how to manage the data; you want to be able to concentrate on the analysis.

Virtually all GIS software hides the algorithms that it uses. It is not difficult to get different results from the same analysis on the same data if you use different packages. With some types of analysis it is possible to check the result on a small area using a subset of the data, but apart from a few academics very few people actually do so. We confess that despite a cynical nature, we take most software on trust.

Software can be obtained to run on any operating system. You do not want to compound your problems by having to learn a new operating system as well.

Be very careful when you consider how easy it is to get data into the GIS. You are going to be spending a lot of time transferring data into and out of the system. Make sure that you do not need to have to keep on adding, say, the class information associated with each colour.

The term *geographic information system* is used extensively. Some packages are little more than mapping or cartographic tools designed to allow you to produce pretty pictures, but not to do any spatial analysis. If you end up constantly exporting your data to do analysis in another package and then re-importing it you will not be happy.

Software packages differ widely in cost. Fortunately the correlation between price and performance is not very strong and you can almost certainly afford something that will do the job. Be careful, however, about the licensing agreements, especially if you are planning to obtain it using an educational discount. The line between research for someone (academic use) and con-stancy (commercial use) is sometimes easy to cross inadvertently.

When you use any GIS package it is advisable to bundle up series of commands in batch files (or shell script or macro) with some comments. There are some people who prefer to get to grips with new software through menu systems, but we do not think they are worth the effort. With a batch file you have a permanent record of how you carried out the analysis: what the input data was, what functions you used and *why* you used those functions (provided you added comments). In conjunction with a README file in each sub-directory (and lots of sub-directories) 90% of the project docu-mentation is automatically there. Even if you do not put comments in your batch files you can at least set them running and go off and do something important – like thinking about what you are trying to achieve and what the next step is.

There are a lot of companies producing software; it is likely that most of them will go 'belly up' at some stage. It is not a question we would pay much attention to, but then we are not very risk-averse with regards to new technol-ogy; neither do we work on many projects that last more than a few years. The criterion is included for those of you with projects that are definitely going to last for decades, such as computerising the records for your local natural history museum.

When you start your GIS project you may have a very limited set of goals. This is only natural. As your knowledge increases and the project progresses you are probably going to do more and more; try and find a package that you can 'grow into'.

GIS software capability can be divided into a number of classes; some classes are unique to GIS, others are more general. Three capabilities fundamental to a GIS are mapping functions, data management functions and spatial analysis functions. More general capabilities include communications, menu and graph-ical user interface development, interfaces with other software packages (e.g. statistical software), file archiving and so on.

Determining your hardware requirements

After the software has been selected it is possible to start considering the hardware. There are three types of hardware:

- equipment for crunching numbers and storing data – PCs and workstations,
- equipment for getting data into the GIS – digitisers and scanners, and
- equipment for producing hard copy – printers and plotters.

At one time your hardware requirements would be driven by the software you wanted to use; nowadays many GIS will run on several platforms. Even a few years ago we would have had to include a long discussion here on mainframes, dumb terminals, workstations and personal computers. Fortunately the distinctions between the different categories are becoming so blurred that it no longer seems very necessary.

With a few caveats our advice on what computer hardware to buy is 'as large and as fast as you can possibly afford'. With a GIS you will quickly get to the stage of wanting to manipulate very large volumes of data, and generally the more memory the computer has the less you will be slowed down. One of the authors has a significant bias towards large UNIX boxes, but will admit (under pressure), that useful work has (probably) been done on 'Macs' and possibly even PCs. You are going to be spending a lot of time in front of your computer, so it is worthwhile getting the biggest, highest-resolution screen you can afford.

As well as the workstation you may also need to consider input and output devices. A small digitising tablet may be needed, but you should be aware that there may be serious copyright implications when digitising (these are discussed in more detail in Chapter 4). In our experience it is not trivial to get digitisers to work properly (or even at all). Digitising tablets vary in size from A4 up to A0, which is big enough for the largest printed maps, but large digitising tables are seriously expensive. Scanners also come in a range of sizes; small desk-top scanners are now quite cheap, but do not forget that an 'image' of a map is not the same as the geographic data.

You will almost certainly need some sort of printer for producing hard copy. How much hard copy you require will depend very much on what sort of project you are working on. Any plotter or printer above A3 paper size will become very expensive and specialised. For A3 and A4, virtually any Postscript colour printer will do. In terms of quality, dye sublimation printers produce close to photographic standard, but thermal wax, laser and ink-jet models can produce high-quality images. Printer technology changes very rapidly. The cost per printed image is an important variable and may differ by an order of magnitude between dye sublimation and ink-jet printers. The problem with giving any advice on printers (and any other computer hardware) is that prices change on an almost daily basis; technology that was exorbitant at the beginning of the year might be the *de facto* standard by the end. Box 3.2 gives some details of different types of printer.

If possible you should ensure that your computer can be networked and linked to the outside world (Internet, World Wide Web, etc.).

Box 3.2   Types of printer

Type of printer	Approximate resolution (dots per inch, dpi)	Relative cost of hardware	Relative cost per image
Dye sublimation	> 400	High	High
Thermal wax	> 400	High	Moderate
Laser	> 300	Moderate	Low
Ink-jet	> 120	Low	Low
Pen plotter[a]	~ 200	High[b]	Low

[a] Pen plotters produce only vector output; the resolution is effectively limited by the width of the finest pen.
[b] Not very popular these days, so relatively expensive.

Determining your work programme

By definition all research projects are unique. Calculating the time and resources needed to carry out unique projects is the bane of our lives: if we knew exactly what we were going to find out and when we were going to find it out, then we would not want to do the project in the first place! Unfortunately you are going to have to make some estimate of how long your project will take.

For a typical project the basic processes you will have to go through are:

1. deciding exactly what you want to do (planning),
2. getting hold of the data (from field work or published sources),
3. importing the data into the GIS (scanning, digitising, typing, electronic transfer, etc.),
4. performing the 'housekeeping' activities (convert to a consistent co-ordinate system, check for errors and omissions, understand the data and generate appropriate keys, colour schemes, look-up tables, etc.),
5. performing the spatial analysis, and
6. presenting the results (design and print maps, tables, graphs, etc.).

The naive among you might expect that the greatest time and effort will be expended on step 5 – sorry! each of stages 2–4 will almost certainly take longer; stages 6 and 1 may also take longer. Obviously the exact balance of time depends on what you are trying to do, the source of data and your experience. As a rule of thumb we would suggest that you budget three-quarters of your time on stages 2–4, 5% of your time for planning and the remaining 20% for analysis and preparing the results. For example, say you are doing a vegetation survey of a nature reserve taking a week to do the field work, then you are probably going to need at least half a day for planning, a week to get the data into the GIS (and make sure the data are coherent), a day to carry out the analysis and a couple of days to produce the output.

These project timings assume that you are reasonably familiar with a particular GIS package. How long is it going to take you to become familiar with a GIS? To become a GIS expert or 'guru' then you probably need to invest several years of daily use; to become professional perhaps six months of regular (daily) use or an MSc course; to become comfortable with a GIS, perhaps twice as long as it took you to become comfortable with the *first* word processor you ever used.

Review

Many GIS projects fail, but technical problems are rarely important reasons for failure; provided that you can clarify in your own mind exactly what you are trying to achieve then you stand every chance of success. There is no shortage of GIS software to suit every pocket and platform; most are increasingly easy to use, but you *must* plan carefully.

Further reading

Independent reviews of existing software are regularly found in the trade and professional journals, like *Mapping Awareness* (but infrequently if ever in the academic journals). The best 'further reading' you can do is find someone who is already using a GIS and see if you can get them to chat about it in an informal manner.

Chapter 4

Sources of data

A GIS is nothing without data. Some input data you will collect for yourself, but there is also a wealth of published data available. In this chapter we first 'lay down the law' about what you can and cannot do with other people's data. We then go on to discuss the three main sources of data: field survey, published data and remote sensing. We finish with a brief recap of sources of error in data and a discussion of some of the practical issues of importing data into your GIS.

Ownership of data

The ease with which information can be copied and reproduced on computer systems, together with the increasing need for universities and research institutes to be 'market aware', has led to increasing interest in maintaining ownership of information and ideas. The safest assumption to make about any data you wish to use is that they will be subject to copyright and intellectual property rights (IPR). Even when information has been transformed or used in another analysis it may still be subject to copyright and IPR. By the same token any data *you* collect will also be covered by copyright and IPR. At the very least you will always need to acknowledge sources of data: whether or not you will have to pay for it is another matter. Licensing agreements vary considerably; for example, it is not unusual to be able to use data for research or educational purposes without charge.

Copyright and Intellectual Property Rights (IPR)

> The first thing we do, let's kill all the lawyers
>
> (King Henry VI, part II, W. Shakespeare)

Copyright is one of the legal 'tools' that can be used to protect intellectual property. Unfortunately, if there is a comprehensible legal definition of just what constitutes intellectual property it has escaped us! Advances in computer technology have made it more and more easy to produce and distribute copies of data and information in such a form that it is impossible to tell whether a particular version is an original or a copy, or even how many copies may have been generated. Because of this, owners of spatial data have attempted to use copyright to protect their interests. Copyright is a *property right* that gives the

---

Box 4.1   Some examples of how copyright might be interpreted

Laws vary from country to country, and neither of us is a lawyer, but a few examples may help to clarify things. If in doubt, ask the copyright holder first.

- If data are made available across a computer network, copies of the data are being made (even if the individual user is unaware of it) and you ought therefore to be licensed to make multiple copies. Viewing data across a computer network can also constitute a *broadcast*, which may be a restrictive act.
- Digital data that are legally considered to constitute a work of *literature* should be able to be viewed as an *artistic* map on a computer screen without infringement of copyright.
- Scanning or digitising is a *restrictive act* because retrieving and storing data in electronic form is an explicit form of copying.

---

owner the exclusive right to perform certain acts (termed *restrictive acts*). The most important of the restrictive acts is the right to copy the work. Others include the right to perform, show or play the work in public and to make an adaptation. Restrictive acts are not exclusive.

The thing that makes copyright law so difficult (and lawyers so rich) is the conflict between two quite reasonable needs. On the one hand the person who put in the hard work, had the bright idea and produced the information should be able to benefit from all that hard work. On the other hand the person buying the information (or data) should be free to enjoy (and use) what they have paid for. Note that, of course, the copyright holder might not be the *author* of the work (in commercial organisations conditions of employment sometimes specify that copyright and IPR of everything produced reside with the organisation such that the employees have *no* rights over their 'own' works!). We can illustrate one of the problems with copyright with an example from this book. For one of the projects we have worked on we used an air photograph to estimate land use. The company charged us £700 for the photograph and then it charged us £800 for an interpreted version (identifying field boundaries, etc.). When we asked if we could use it as an image on the front cover it said: 'fine that will be £1500 plus 17.5% tax please', so we said 'no thanks'.

Larner (1992) provides a detailed discussion about whether an act as simple as viewing a map on a screen constitutes copying (for which a royalty could be charged) or whether it is a legitimate use. Our understanding is that there have not yet been enough court cases to decide exactly where the boundaries lie between copying and legitimate use. Box 4.1 provides some examples of how copyright might be interpreted.

Just because something is supplied without an explicit reference to copyright you cannot assume that you are free to use it in any way you like. Because of the difficulty in proving that copyright has been infringed when data has been 'reformatted' or manipulated it is not unknown for publishers to include fictitious entries in encyclopaedias and biographies – there have been several embarrassing (and amusing to the rest of us) instances of experts attributing pieces of music and works of art to fictitious characters included to catch people infringing copyright. There has been more reticence about including fake geographic entities on maps, but on small-scale **thematic maps** deliberate inclusion of oddities has been known: Monmonier (1991), for example, provides some examples of non-existent towns.

GIS software can often be obtained for educational or research purposes at reduced cost (or occasionally for free). In the UK a body called the Combined Higher Education Software Trust (CHEST) exists to negotiate reduced rates for all types of software. Where life becomes difficult is where the legal boundary between doing research and being a consultant becomes blurred. It is becoming increasingly common to consider that the software is free and that what is being purchased is the right to *use* the software for specific purposes.

Field collection of data

It seems that the ecologist is never happier than when he or she is out on field work. The attribute data used to populate your GIS can take any form. The two primary ways to record data in the field are graphical and tabular. In either case it will be necessary to carry out some sort of transformation of the data before it can be used in the GIS.

This is not the place to go into too much detail about ecological field work; however, a few points need reiterating: plan, test, check.

The first thing to do is plan the work: decide what you want to achieve, do a reconnaissance, ask for advice and if possible do a test run. Field work is most successful when it is planned to test a clear hypothesis (as the data comes in you may very well find new questions to ask but at least start with an idea). It is always best to check with a statistician before you start designing your 'experiment'; it is also worthwhile to check with a GIS expert that the spatial information you plan to collect is going to be sufficient. The main expense of any field work is getting the ecologist to the study site. Once you are at the site the extra cost of another datum item is minimal.

Always check the data you have collected as you collect them; especially, always try to abstract them the same day you collect them. Never leave checking the data and quality assurance to the end of the week or, heaven forbid, the end of the season. Even if you do recognise a gross error after a week or a month it is too late to do anything about it. As a salutary example: later in this book there are some data from a study of the distribution of orchids at Toternhoe. The electronic theodolite we used to collect the data always presented the reduced co-ordinates as northings, then eastings (rather than using the conventional order). Up until lunch we reversed the order so they were the right way round as we did the recording, but after lunch there was a page of results where Richard forgot! Because we reduced the data at the end of the day we spotted a 'funny' – the data seemed to suggest that my colleague (with the prism) spent some time walking over the same part of the site after lunch as she did before lunch. So we checked the field notes, realised the mistake and so no harm was done. Could you remember what order you walked across a site at the end of the week or the end of the month? Even when you are doing something as simple as recording a species list as someone inspects a quadrat it is always valuable to check before you move on that you have seen what you expect to see. In the past one of us spent some time recording for a botanist working on a heather moorland (health and safety regulations meant he could not work alone). Anyway Richard could not tell you the difference between *Sphagnum recurvum* and *Sphagnum cuspidatum* if they jumped up and bit him, but you do not have to record too many quadrats before you can guess what to expect and ask 'what no "spag rec" in this one?' It takes only seconds to ask, to check and to move on, but if you do not check at that instant it is too late – who knows what increases and declines in species are based on one observer being in more of a hurry than another?

There are four forms that spatial data from field work can take: point recordings, quadrats, transects and maps. In all cases the most important feature is the accuracy of the locations, because without an accurate location your data will at best be meaningless and at worst misleading:

- location of point records
- location, orientation and size of quadrats
- location, orientation and size of transects
- scale, resolution and reliability of base map.

Most ecological data are collected using some form of ground survey. To locate where the samples or observations were taken, the choice lies between land survey systems and GPS.

Field mapping

The most common form of graphical data collection is to annotate an existing base map in the field. Given free access to the land, detailed large-scale mapping and simple data requirements (such as land cover) it may be sufficient to delineate boundaries and areas by 'eye' (sketch areas in). Where detailed base maps do not exist it may be necessary to make more or less detailed measurements using tape measures or theodolites or GPS. Where the requirement for data is much more detailed than just simple land cover, additional data notes will have to be made.

All land survey systems are based on measurements from defined control points. It is often necessary to establish 'temporary' control points in the study site from which sample points can be fixed by either distance or angle measurements. Standard modern land survey is carried out using a 'total station' that combines the functionality of a theodolite and an electronic distance measurement (EDM). With modern equipment angular measurements can typically be made to 1 second of **arc** and distances measured to a few parts per million. Such precision is usually well beyond what is needed for ecological studies; however, there is little if anything to be gained from measuring at a lower precision with such equipment.

Global positioning systems

An increasingly common method for locating your position without the need for extensive survey is the global positioning system (GPS). A GPS works by interpreting the signals coming from a swarm of 21 satellites (plus three spares). Each of the satellites contains a very accurate clock and emits a unique signal that includes information about the time; if you receive the signal at a known instant you could estimate how far away it is. As the position of the satellite is known at any point in time the GPS receiver must be located somewhere on a circle centred on the point immediately under the satellite. If you have the distance from two satellites the receiver can be in one of two points (where the two circles intersect), and distances to three satellites will provide a unique location. In practice observations from a fourth satellite are used to check the times. An estimate of your location can usually be obtained within a few moments. Unfortunately, for reasons best known to themselves, the owner of the satellites (the US military) deliberately degrades the signal emitted by the satellites so that instead of having an accuracy of centimetres (which would allow you to relocate a marker peg easily) it is typically in the order of many tens of metres. Of course if you were willing to sit around for long enough and take enough measurements the mean value would become increasingly accurate. A more realistic alternative is to employ a differential technique. Differential techniques typically make use of two receivers: one remains at a fixed point while the other moves around the area of interest. At the fixed point the location estimated by the GPS is recorded at frequent regular intervals, and these apparent changes in location of the fixed point are then used to correct the location of the roving GPS receiver. It is necessary to ensure that both receivers are using the same satellites to estimate their position at the same instant, as each satellite has a different random error. Note also that vertical 'errors' can be considerable, geoid–spheroid separation can exceed 100 m.

Published data sources

There are many sources of spatial information, ranging from the free and co-operative to the expensive and confidential.

Where prepared or commercial sources of information exist they may offer several potential advantages over trying to capture information yourself. These are summarised in Box 4.2. Of the advantages listed, the first has the most to recommend it with respect to research purposes.

Prepared electronic data sets are supplied by many national mapping agencies, such as the Ordnance Survey in the UK and the Geological Survey in the USA, as well as international co-ordinating bodies such as the United Nations Global Resource Information Database (UNEP–GRID) or the Comité Européen de la Cartographie (CERCO). In addition, many government departments and non-government organisations prepare and release information.

Box 4.2   Potential advantages of purchasing data

1. Conformity: other people will be using the same data so that your analysis and results should be open to interpretation by a wide audience.
2. Ease: you do not have to do any digitising or field work to collect the data, though you will *still* need to check the data.
3. Avoidance of decisions: consideration of what categories to use, how to define them, what is an appropriate level of precision, and so on, can be left to others (hopefully experts!).
4. Cost: buying data is often cheaper than going out and collecting it yourself (but some monopoly suppliers operate what appear to be discriminatory pricing policies).
5. Time: it is almost always quicker to buy than to collect.
6. History: for some studies, notably those with a historical dimension, prepared data sets may be the only available source of information.

There is always a temptation to collect (and store for interactive use) as much data as possible. Certainly, we have never knowingly refused a data set (even if we had *no idea* what we might wish to do with it at the time). However, large data sets can pose a number of problems, not least of which is the need for efficient storage. This often requires data sets to be broken down, whether spatially or thematically, into manageable sub-sets, which require careful management and documentation. Really large data sets also slow computers down, and may overload them altogether. Potential problems with spatial data sets have already been discussed (Chapter 2). However, it is worth reiterating that you can *never* take data on trust. It is imperative that the characteristics of the data set, and possible problems with it, are known. It might be worthwhile to ask other GIS users what their favourite *bête noir* is. Our own particular favourites are maps using unorthodox projections (and more especially orthodox projections with unorthodox parameters), maps that turn out to be compilations of information collected in subtly different ways from different regions, and digital maps that turn out to be scanned images of printed maps.

'Conventional' sources of 'published' data

Different countries have different attitudes towards national data sets. For example, the UK has many detailed, national data sets collected and revised over many years, available in various electronic formats as well as on paper. Government-funded institutes and departments concerned with soils, geology, river flow, weather, agricultural land use, topography and land cover all sell their data and some of the data cost a lot of money. Other valuable data sets are completely unavailable; for example in the UK the Ministry of Agriculture requires farmers to record the crops grown in every single field in England and Wales. That information is collected and stored on a GIS, but nobody but the people who check on the agricultural subsidy payments are allowed access to it. Other countries, such as the USA, have much more liberal attitudes to charging for data. In less developed countries national data may be free or for sale, but may be rather elderly.

If suitable published data exist then there are a variety of data capture methods that exist. Vector data will usually be captured with some form of digitising and raster vector by **image processing** or scanning. Increasingly data will be available in one of a wide variety of electronic formats. Data formats used for GIS are or should be designed to preserve all the attributes of the data, but unfortunately some confusion exists with what are effectively graphics formats. Whatever the format of the data you must always make yourself aware of the source of the data and what the data were collected for.

The most common form of published data used to populate a GIS are a map or plan. There are many types of map: **cadastral**, chromatic, compiled, composite, index, **orthophoto**, photo-mosaic, planimetric, plastic relief, residual, thematic, topographic, trend surface. There are a few things you should realise about a map before you start to use it. The first and most important is that a map is a model of the world, it is a cartographic invention. Cartographic conventions are used to simplify and codify spatial information to present a coherent, comprehensible view of some aspect of the real world. Each type of map is designed to communicate a particular view or aspect of reality: topographic, soils, geological, weather, vegetation, population, road, ownership (cadastral) and so on. Maps now come in two forms: printed and electronic.

Where data exist as printed maps it is important to know which projection and **datum** were used. For example, in the USA large-scale mapping is most commonly on either the NAD27 datum or NAD83 datum. While the parameters of the two datums are very close, they can give rise to northing shifts ranging from +192 m in Los Angeles to +223 m in Boston; easting shifts can vary from +47 m in Boston to −97 m in Los Angeles (Welch and Homsey, 1997).

National mapping

Every country has a national mapping agency; however, the amount of information that they have in digital format varies. In the UK the Ordnance Survey (OS) is very active in making its information available in digital format (though it is often very expensive). Table 4.1 shows some of the electronic information that is available from the OS (in June 1996).

OS data are supplied in tiles varying from 1×1 km (LandLine) to 100×100 km (BaseData.GB) for an annual fee (plus data-use royalty, plus a hard copy royalty fee). In 1997 the most expensive data were about £56 year^{-1} km^{-2} (LandLine); the cheapest, £105 year^{-1} per 100×100 km tile (BaseData.GB). All OS products are updated after a significant amount of change. One unit of change is approximately equivalent to a single suburban home, and a 'significant change' is more than 20 units; the OS will (for a premium) supply annual updates of maps that have changed by more than one unit.

Table 4.1 Examples of electronic data from a national mapping agency: the UK's
Ordnance Survey

Product	Scale	Coverage/description
LandLine	1 : 1250	Urban areas/vector data in 57 categories plus six layers of text
	1 : 2500	Rural areas/vector data in 57 categories plus six layers of text
	1 : 10 000	Whole country vector data in 57 categories plus six layers of text
Meridian	1 : 50 000	Whole country/vector data in 26 categories
Strategi	1 : 250 000	Whole country/vector data; 118 feature codes in six layers
BaseData.GB	1 : 625 000	Whole country/vector data; 88 feature codes in six layers
Land-Form Profile	1 : 10 000	Whole country/contours or 10 m raster grid
Land-Form Panorama	1 : 50 000	Whole country/contours or 50 m raster grid
B&W raster	1 : 10 000	Whole country/scanned image (400 dpi) of printed 1 : 10 000 map
Colour raster	1 : 50 000	Whole country/scanned image (254 dpi) of printed 1 : 50 000 map

All the products referred to in Table 4.1, except Meridian, are based directly on published maps. Although they can be used in a GIS, the cartographic heritage of the products should always be borne in mind. As well as
the products shown in the table, the OS also produces other electronic data
sets that are less likely to be of interest to ecologists. Other OS data sets
include Address-Point (location of all 25 million postal address in the UK to
within 0.1 m), ED-Line (boundaries of the enumeration districts used by the
Office of Population and Census), Boundary-Line (all administrative boundaries), SABE (all administrative boundaries in Europe) and Oscar (structured
road network at various levels of complexity).

In the USA the national mapping authority is the Geological Survey, which
offers a wide range of products: the largest national mapping is at 1 : 24 000
scale (7.5 minute quads), with other major map series at 1 : 100 000 and
1 : 250 000 scales. A variety of interesting electronic products exist, including:

- digital terrain models,
- orthophotographs (geometrically corrected air photos annotated with printed line work and text),
- digital line graphs, and
- digital raster graphs.

The extent of the coverage of the different products can be checked through the Internet (**http://mapping.usgs.gov/www/products/status.html**) (USGS Earth Science Information Center 1-800-USA-MAPS). The USGS also operates a number of regional centres like the Rocky Mountain Mapping Center in Denver.

Published ecological and environmental data

Most countries have considerable amounts of environmental data collected by a variety of government organisations and institutes. In our own ecological research in the UK we make considerable use of data on land cover, topography, soils, geology and species distribution. Table 4.2 provides a brief description of some of these commonly used environmental data sets. Note that details of costs are rather vague for some of these data sets – welcome to the wonderful world of negotiation! Appendix A provides contacts for these data sets.

In the USA the Geological Survey (responsible for national mapping) also has a Biological Resources Division (Biological Resources Division, USGS, US Dept of Interior, Office of Public Affairs, 12201 Sunrise Valley Drive, Reston, VA 20192). Data on species and land cover, especially on the publicly owned lands, can be obtained from them (**http://biology.usgs.gov/pub_aff/natprog.html**). As well as the national data sets many states seem to operate their own GIS server. For example, you can download data on land cover, land use, hydrography, historical population for Connecticut in ArcInfo and MapInfo formats straight off the Internet (**http://magic/lib.uconn.edu**); you can get similar (but different) data from places like South Carolina (**http://www.dnr.state.sc.us**), San Diego (**http://www.sandag.cog.ca.us**) or the Great Lakes (**http://www.cciw.ca/gl**).

'Unconventional' sources of 'published' data: the Internet and World Wide Web

There is increasing interest in the Internet and the World Wide Web (WWW) as sources of information and resources. Although most of the Internet resembles a pile of junk mail there are significant resources for ecologists to exploit; for example:

● What techniques to use?
● What software and hardware are available?
● What data exist?

A word of warning about the Internet; 'surfing' can be a big time waster. Try to remain focused on the task in hand and avoid getting involved in futile discussions! With a system as large and as fluid as the Internet it would be pointless to attempt to describe or document it in detail. There are so many guides available that we do not even feel confident about which are the best to recommend. It should be borne in mind that, with a very few exceptions,

Table 4.2 National environmental/ecological data sets in electronic format for the UK

Data set/owner	Description
Land cover/Institute of Terrestrial Ecology	Estimate of land cover in 25 classes at a resolution of 25 m. Derived from Landsat TM summer and winter scenes captured around 1990±2 years
Species distributions/Institute of Terrestrial Ecology	Distribution of all vascular plants and most lower plants, many groups of invertebrates (especially, ground beetles, butterflies and other high-profile species)
Digital terrain model/Institute of Hydrology	Ground elevation on a 50 m grid – 'corrected' so that rivers always run downhill
Precipitation, evaporation and river flow/ Institute of Hydrology	Maps at 1 km resolution of monthly rainfall, average annual rainfall, maximum expected rainfall, evapo-transpiration, winter rainfall acceptance Time series of gauged flows (daily mean flows)
Human census data/Manchester Information Datasets and Associated Services (MIDAS)	Twenty-nine variables from the 1981 and 1991 population census have been **extrapolated** to 200 m grid. Variables include population (classed by age and gender), households (owned, rented), car ownership. Other census data available but need to be extrapolated
Soils/Soil Survey and Land Research Centre	Digital data scanned from 1 : 250 000 scale mapping. Provides the dominant soil association in an area Hydrology of soil types (HOST) classes associations into one of 26 types based on percolation and probable underlying geology
Solid geology/British Geological Service (BGS)	Solid geology: only the 1 : 250 000 is available for the whole country. Drift geology exists only on 1 : 50 000 map sheets, but coverage is not complete
Agricultural land use/Ministry of Agriculture, Fisheries and Food (MAFF)	The June census records crop areas and livestock numbers for all full-time farms. Small-area statistics are published for amalgamated parishes (before 1988 parish totals were available). Paper records go back to 1866; digital version available for recent years

information available on the 'net' is not refereed or subject to reliable quality assurance mechanisms. Those Web pages and Internet resources mentioned in this book can only safely be regarded as sources of personal idiosyncrasy and anecdote: they should not be considered to represent accepted fact, majority opinion, endorsements or reliable recommendations. There are various guides to 'good etiquette' on the Internet, but as a general rule, using common sense should suffice. Most long-term users are tolerant of 'newbies', although, by a strange coincidence, tempers in the more academic corners of the 'net' tend to flare every October.

Internet resources fall into several categories. In our professional capacity we make use of the following (in descending order of frequency):

- e-mail (the best way to communicate ever devised),
- newsgroups,
- databases, and
- Web pages.

The main advantage of e-mail is the ability to communicate quickly and efficiently with fellow researchers, students and practitioners all over the world without worrying about time zones, and it has done a huge amount to open up international dialogue between ecologists.

Newsgroups exist to discuss a bewildering array of topics. Many commercial GIS producers have one or more newsgroups where you can ask technical questions ranging from the blindingly obvious to the impossibly obtuse. For example, a GIS such as GRASS has a programmers' newsgroup for people wanting to write modules ('info.grass.programmer') as well as a users' newsgroup ('info.grass.user'). There are also newsgroups and discussion groups relating to the use of software and for a variety of ecological applications. For example, there are newsgroups to discuss GIS and biological conservation (**consgis@uriacc.uri.edu**) and GIS and coastal and ocean research (**sea-gis@listsrv.hea.ie**). At least 17 newsgroups exist discussing scientific aspects of biology (**sci.bio**) including groups concerned with botany, conservation, entomology, homoptera, ethology and phytopathology. Other 'ecological' newsgroups exist under different headings. Some groups generate dozens of messages a day, others only one or two a month. Responses to questions can normally be obtained within a few hours, but take the time to read the 'frequently asked questions' (FAQ) to avoid unnecessary duplication. Box 4.3 lists newsgroups of potential interest to ecologists.

GIS software producers often have Web pages where they announce new developments and generally showcase their products. A couple of Web pages that an ecologist might be interested in are the GRASS Web page, which includes a couple of ecological examples, and the ESRI home page.

Interactive GIS over the Internet is also possible, although often very slow. An interesting system for producing maps of different parts of the world using a variety of projections can be found at **http://137.166.132.18/cgu-bin/gis/Map**.

There are some sources of free spatial data on the 'net' ranging from real-time images generated by weather satellites to images scanned from printed

---

Box 4.3   A 'taster' of newsgroups of potential interest to ecologists

geo-computer-models@mailbase.ac.uk – modelling spatial processes
esri-l@esri.com – ESRI products (ArcInfo, ArcView, etc.)
comp.soft.sys.gis.esri – alternative list for ESRI products
info.grass.user – for users of GRASS
Idrisi-l@toe.towson.edu – for users of Idrisi
mapinfo-l@csn.org – for users of MapInfo
consgis@uriacc.uri.edu – GIS for biological conservation
sea-gis@listserv.hea.ie – GIS in marine and coastal studies
coastgis@irlearn.bitnet – GIS for coastal studies
acdgis-l@awiimc12.bitnet – GIS discussion mostly in German
AfricaGIS (e-mail **listserve@orstom.fr**) discussion mostly in French

Newsgroups where spatial ideas may be more peripheral
bionet.biology.computational – numerate biology
aliens-l@indaba.iucn.org – invasive plant species
sci.bio.entomology.misc – entomology
sci.bio.conservation – conservation (can be quite 'political')
sci.bio.ecology – all sorts!

---

atlases. Unfortunately many of these are either global data sets with very coarse resolution or are restricted to the USA. However, the location of UK and other European data sets can often be found on the 'net'. When you have located a data set of interest, transferring data is relatively straightforward and convenient: certainly more convenient than transferring it using physical media such as tapes or disks or, heaven forbid, paper. Box 4.4 lists some Web pages of potential interest to ecologists (but there is no guarantee that these still exist).

Remote sensing

Remote sensing includes a variety of techniques for collecting the electromagnetic information reflected from the ground by instruments mounted on satellites, aircraft or on the ground. Data can be stored chemically (as photographs) or electronically. Most techniques produce recognisable images, although some such as synthetic aperture radar are virtually impossible to decipher by eye. The basic division in techniques is between passive systems (which rely on energy from the Sun that is reflected from the object) and active systems (which emit energy that is then reflected back by the object). Box 4.5 provides a glossary of some of the more commonly used terms.

As a tool for collecting data remote sensing has the capability to provide synoptic views over very large areas very quickly. Within ecology remote sensing is increasingly being used for a wide variety of tasks. The Land Cover

Box 4.4 Potentially interesting Web pages	
http://www.geo.ed.ac.uk/home/ giswww.html	List at the University of Edinburgh with links to hundreds of GIS sites
http://www.cecer.army.mil/ grass/viz/VIZ.html	GRASS home page
http://www.esri.com	ESRI home page – ArcInfo, ArcView, etc.
http://gopher.gis.umn.edu/11/ rdgis/lists/esri-l	Archive of thousands (tens of thousands?) of questions about GIS (using ESRI products)
http://www.idrisi.clarku.edu	Idrisi Project home page (regional centres offer some information in the local languages)
http://www.spotimage.fr francais/use/envur/ ue_env.htm	examples of SPOT satellite data
http://earthl.esrin.eas.it	European Space Agency – including searchable index of images
http://rsd.gsfc.nasa.gov/rsd	NASA – public use of remotely sensed data
http://137.166.132.18/cgu-bin/ gis/Map	Interactively try out different types of projections
http://www.regis.berkeley.edu/ grasslinks/index.html	Interactive mapping of San Francisco Bay, North Bay Protection Plan, etc.
http://plue.sedac.ciesin.org/ plue/ddviewer	Interactive mapping (US census data)
http://acorn.educ.nottingham, ac,uk/ShellCent/maps	Studies on making maps easy to read

Map of Great Britain has been described above and has been used in a wide variety of environmental and ecological studies. Synthetic aperture radar (SAR) has been used to estimate forest biomass, flooding under trees and limits of freezing conditions. Light detecting and ranging (LIDAR) appears to have great potential for describing micro-topographic changes and linear features. Data on **reflectance** in the near-infrared help describe the health of vegetation and the presence of sediment and plankton in water bodies (see Kasischke *et al.*, 1997, for a recent review of imaging radars in ecology).

Remote sensing provides an image at a specific point in time, so when used to describe change it is important to be able to determine exactly when the

Box 4.5   Glossary of remote sensing terms

angular resolution – a measure of the smallest angular separation between two objects – the smallest object that can be seen

band – a defined interval in the electromagnetic spectrum, (see also *panchromatic* and *multispectral*)

change detection – comparison of two images of the same area acquired at different times (a surprisingly difficult operation using satellite data)

image enhancement – improving the visual appearance of an image

image processing – turning the image into information

multispectral – measuring the scene over a number of distinct narrow *bands* in the electromagnetic spectrum. Typical sensors use three or four *bands*, but can exceed 100

orthorectification – process of geometrically correcting an air photo

panchromatic – measuring the reflectance over a wide *band* of the spectrum; typically the range will be close to that seen by the human eye

photogrammetry – process of extracting geometric (shape, size) information from photographs

pixel – contraction of 'picture element': smallest discrete element on a satellite image; used loosely in GIS to indicate an individual cell in a raster grid or matrix

rectification – process of aligning an image to a particular map projection and removing distortions due to the imaging system, viewing angle

reflectance – ratio of the energy reflected by a body to that incident on it; depends on wavelength and direction

registration – process of aligning the image with another image or map

spectral signature – a unique pattern of reflectance resulting from incident energy with a range of wavelengths. Central to the idea of distinguishing different objects from how they reflect or absorb electromagnetic radiation

stereo imaging – two images of the same area taken from slightly different points so that the illusion of a three-dimensional object can be recreated

track – or ground track; vertical projection of the flight path

images were collected relative to specific events and time of year, There are many types of remote sensing, which differ in the type of platform (satellite, aircraft, etc.), spectrum sampled and effective ground resolution.

Passive sensors detect the reflection of the Sun's energy from the surface and typically work in the visible and infrared part of the spectrum. Active systems typically transmit radio waves (that is radar) or use low-powered lasers. Data may be collected by the sensor using photo-chemistry (that is photographs) or by photo-electrical means (charged-coupled devices). For the same elevation and viewing angle, photographs provide much higher resolution than electronic images but require much more human intervention to interpret. In the majority of cases images taken with the device pointing straight down are the most useful although oblique images can be used. A remotely sensed image will have to be registered to some co-ordinate system before it can be used with other data. Images captured on air photographs are usually fairly easy to register, as small features on the ground are easily identified; these points can then be used to reconstruct the orientation of the camera as the image was taken. When a scanning instrument is used on an aircraft it is difficult to register the image for two reasons, one is that each scan line will have a slightly different orientation from the previous one and second because the resolution is often too coarse to allow easy identification of ground points. Some experimental work has been carried out using sensors that record a scene instantaneously by using a matrix of sensors rather than using a single line of sensors or a scanning point. Satellite images are also often difficult to register, but fortunately this is usually done by the organisation supplying the images. Registration of satellite raster images may be expected to be within one or two pixels of the correct position.

It can be quite difficult to visualise quite what the different resolutions of satellite data mean. As a 'rule of thumb', 5 m resolution corresponds roughly to 1 : 10 000 scale mapping; 30 m resolution (say Landsat Thematic Mapper) is approximately equivalent to the level of detail you would get on 1 : 100 000 scale mapping (note of course that most 1 : 100 000 scale maps are cartographic products that include many objects that have been exaggerated so they can be seen – such as most roads, railway lines, canals and so on).

Terrestrial remote sensing

Terrestrial-based cameras and sensors do not appear to be widely used in ecology. We have been trying to persuade some of our colleges to use terrestrial **photogrammetry** to measure the fine details of ripples on a muddy beach. We have another scheme to use a camera on a long pole to get data to describe the canopy of dwarf shrubs on a moorland. Unfortunately the mathematics of producing geometrically correct data from sensors looking horizontally (or nearly horizontally) are not trivial and the equipment to do so is not widely available, so both schemes remain as ideas.

Air photography

Remote sensing in the form of photographs has a long history: almost as soon as planes were invented people started taking ariel photographs. Cameras designed for air photography have a large format (the negatives are 6 inches across) and have very complex lens systems to ensure that the image is geometrically correct. Complete coverage of the UK exists, and many counties have complete coverage at five to ten-year intervals since the 1940s. In the USA the National Aerial Photography Program has completed two cycles (1987 to 1991 and 1992 to 1996) and the current cycle is due to be completed in 2002. The resolution of an air photograph is primarily controlled by the height of the aircraft and the focal length of the camera; difference in grain size on different emulsions is not usually critical in defining the resolution. As

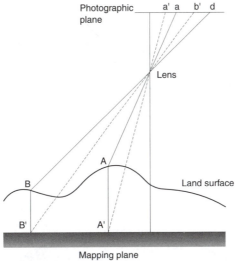

Figure 4.1 Air photographs and planimetric position.

a rough guide black-and-white air photography at 1 : 10 000 scale allows you to interpret (or identify) all features you might be expected to identify at a distance of just under a kilometre on the ground (Kilford, 1970).

Air photographs are typically taken with a 60% fore-and-aft overlap and a 15–20% side overlap to ensure that even with roll, pitch and yaw it will be possible to construct a geometrically correct stereoscopic image. The emulsion used on photographs may be chosen to record different parts of the spectrum. When recorded electronically it is usual for a number of fairly narrow spectral **bands** to be used; these bands are often selected to help identify particular types of feature, for example use of a near-infrared band for detecting chlorophyll.

Emulsions used in photography are typically sensitive in the visible and near-infrared range of the spectrum, when developed images are usually either grey-scale, false colour or true colour. Maximum detail can be detected by human interpreters on grey-scale images, but interpretation is easier on true colour images. False colour is needed to record near-infrared images and is often useful in studies of vegetation.

Until recently geometrically correct stereo images could only be constructed and measured in large complex optical-mechanical devices. Now, however, software has been developed that enables measurement to be taken within computer software. The initial look at an air photograph can seduce you into thinking that all you need to do is 'paste' it onto your GIS as a backdrop and away you go; unfortunately, life is not so simple. There are three differences between the perspective nature of an air photograph (or satellite image) and the 'correct' mapping plane: topography, Earth curvature and atmosphere. Topographic distortion is shown in Figure 4.1, the distance A–B is estimated from the location of a and b on the photographic plate, the correct planimetric position for the points would be A'B' corresponding to a'b' on the plate. Note

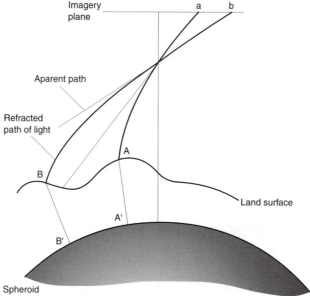

Figure 4.2   Effect of atmospheric distortion and the Earth's curvature.

that the amount of distortion depends on the focal length of the camera, the height of the camera above the ground and the amount of changes in relief. Also note that the 'correct' planimetric distance A′–B′ on the map does not correspond to the distance A–B across the land surface or to the straight-line distance.

Atmospheric distortion arises from the fact that with height atmospheric pressure drops and the refractive index of the atmosphere changes. If the altitude is high enough for the atmosphere to distort the path of light then the curvature of the Earth may also be important. These effects are illustrated in Figure 4.2.

By its very nature all air photography has a slight tilt to it (the aeroplane will be pitching or rolling to some small extent). Even a small amount of tilt changes the geometry of the photography. A major part of photogrammetry is trying to recreate the exact geometry that existed when the photograph was taken. The effect of a severely oblique photograph is shown in Figure 4.3.

LIDAR

LIDAR is a new approach using scanning lasers when mounted on an aircraft to measure elevation. LIDAR stands for light detecting and ranging. The only LIDAR data we have managed to get hold of so far is a 2×3 km block in the south of England – which is very impressive. Elevation data are recorded to the nearest *millimetre* with a data point recorded every 3–4 m across the surface. At that level of detail individual hedges and ditches can be detected. We think that it has the potential to assess the quality of woodland and scrub and

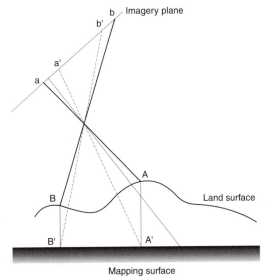

Figure 4.3 Oblique photography.

differentiate between woodland and forest, but it will be a while before we know if it is going to be as valuable as we think it is going to be.

Multispectral scanning instruments

Most multispectral instruments are mounted on satellites, but over the last three or four years they have become more common as aircraft-based systems. Satellite data can be obtained from a variety of sources. Imagery from the Landsat and SPOT satellites is commercially available and often used for ecological studies. Information about commercially available imagery can be found on the 'Internet'; Table 4.3 provides some of the main Web pages (extracted from Plumb *et al.*, 1997 and Danson and Plummer, 1995).

Satellite data now extend over a considerable period of time, in the case of Landsat over decades. As a time series Landsat compares very well with many 'long-term' ecological studies. Peterken and Backmeroff (1988) found a 'half-life' of less than ten years for monitoring schemes set up as woodland nature reserves; that is, in less than a decade more than half had been lost or destroyed, the researcher had lost interest or something else had happened.

Whole books are written about remote sensing and the principles of image analysis; an interested reader is referred to Lillesand and Kiefer (1987). Over recent years the differences between image analysis and GIS are becoming increasingly blurred; image analysis software now often offers some GIS and spatial analysis, and GIS software offers some image analysis and classification functionality. Remote sensing data are really only useable with GIS software that can manipulate raster information; without raster capability it can only be used as a pretty backdrop. Data captured electronically are almost always

Table 4.3 Sources of information on remote sensing images on the Internet

Company and country	Mission	Web page
EarthWatch, USA	EarlyBird* and QuickBird	http://www.digitalglobe.com
Indian Space Research Organisation, India	IRS-1C and IRS-1D	http://www.gaf.de/euromap http://www.spaceimage.com
Orbimage, USA	Orbview-3	http://www.orbimage.com
Space Imaging, USA	Carterra-1	http://spaceimage.com
SPOT Image, France	SPOT	http://www.spot.com
Priroda (Worldmap), Russia	Resours-F1 Resours-F2 Resours-F3	http://www.augusta.co.uk/tentoten http://www.augusta.co.uk/worldmap http://www.eurimage.it/Products/ Prod_Servs.html
SPIN-2, Russia	KOSMOS	http://www.augusta.co.uk/tentoten http://www.eurimage.it/Products/ Prod_Servs.html http://www.coresw.com http://www.spin-2.com
NSPO, China	Rocsat-3	http://www.nspo.gov.tw
German Space Agency, Germany	MOMS-2P	http://www.nz.dlr.de/moms2p/ index.html
NASDA, Japan	AVNIR* and AVNIR-2	http://mentor.eorc.nasda.go.jp/ ADEOS http://alos.nasda.go.jp:80/alos/ what.html
NASA, USA	Clark and Lewis*	http://www.crsphome.ssc.nasa.gov/ ssti/trw/trwintro.htm http://www.crsphome.ssc.nasa.gov/ ssti/cta/ctaintro.htm
	Corona and Lanyard	http://edcwww.cr.usgs.gov/glis/hyper/ guide/disp http://edcwww.cr.usgs.gov
	EOS AM-2	http://eos-am.gsfc.nasa.gov

* Missions failed!

classified before they are used. Classification is an important exercise and should not be treated lightly.

Spatial resolution of satellite remote sensing refers to how much ground is shown on a single pixel, but although pixels are always square the length of just one side is usually specified. Pixel size varies from 1 km to 80 m for MSS AVHRR data to 30 m for Landsat Thematic Mapper, 10 m and 5 m for SPOT sensors and even finer for some of the sensors just coming on line (Table 4.4). Spy cameras and sensors are known to produce very high resolution, but the coverage is more limited (of course if you are interested in land use around military airfields or Washington DC you may get lucky!). Temporal resolution varies from a few times per year (for spy satellite cameras) to a few days. Table 4.4 shows some of the characteristics of existing and planned sensors with their availability, sensors and resolution (Danson and Plummer, 1995).

The spectral response is the cumulative effect of all the surfaces within the pixel; even at moderate spatial resolution, such as Landsat, there may be several very different land cover types within a single pixel. For example, a single Landsat pixel might include part of the roof of your house, grass, trees and the pond in your garden and the car in the driveway; a Boolean classification will reduce the information to a *single* class. Spectral resolution refers to the width of the wavebands and their positions in the electromagnetic spectrum. It stands to reason that the more finely you divide up the spectrum the greater the *spectral resolution* is, therefore a sensor that divides the spectrum up into 200+ slices has a higher spectral response than one that divides it up into just four slices; however, not all parts of the spectrum are equally informative.

An issue that is usually glossed over is bias in the sensor. Within the pixel the sensor may be free of bias or have a centre bias that effectively gives more weight to the features in the centre of the field of view or there may be other geometric bias. In addition to geometric bias there may be overlap between adjacent pixels so that reflectance from the same point will contribute to more than one pixel. Alternatively there may be points that do not contribute to any pixel.

A satellite remote sensing image is rarely an exact analogue to the colours that the human eye sees. To understand a remote sensing image it may be helpful to think in terms of 'why is this area red?' rather than 'what are those red bits?' The incident radiation received at the sensor is made up of radiation reflected from the surface minus what is absorbed by gases and particles in the atmosphere, plus radiation scattered and re-emitted by the atmosphere. Correcting for the effect of atmosphere is an important aspect of making sense of remote sensing data. In some areas clouds and haze obscure the ground for considerable amounts of time. Because of the size of the satellite image geometric correction for the curvature of the Earth has to be applied.

So what happens when sunlight hits a patch of vegetation? In the visible portion of the spectrum, blue light and red light are absorbed by plant pigments; carotenes and xanthophylls affecting blue light, chlorophyll absorbing red light; the more chlorophyll there is the lower the amount of reflected radiation, green light is absorbed less (hence plants look green). In the near-infrared,

Table 4.4 Some current and planned remote sensing instruments

Organisation and country	Mission	Availability	Sensor(s)	Resolution (m)
NASA, USA	Landsat[a]	1972 to date	Multispectral	30
EarthWatch, USA	EarlyBird[c]	Early 1997	Panchromatic	3
			Multispectral	15
EarthWatch, USA	QuickBird	End 1998	Panchromatic	0.82
			Multispectral	3.28
Indian Space Research (ISRO)	IRS-1C	End 1995	Panchromatic	5.8
	IRS-1D	1997	Panchromatic	5.8
ORBIMAGE, USA	Orbview-3	Mid 1998	Panchromatic	1–2
			Multispectral	4
Space Image, USA	CARTERRA-1	Late 1997	Panchromatic	1
			Multispectral	4
SpotImage, France	SPOT 4	Early 1998	Panchromatic	10
	SPOT 5	2002	Panchromatic	5
			Multispectral	10
Priroda, Russia	Resours-F1[b]	1974–1993	Panchromatic	5
	Resours-F2[b]	1988–1993	Panchromatic	8
			Multispectral	10
Worldmap, Russia	Resours-F3[b]	1978–1993	Panchromatic	2
SPIN-2, Russia	KOSMOS KVR-1000[b]	1983–1993	Panchromatic	2
	KOSMOS TK-350[b]	1983–1993	Panchromatic	8–10
	MIR-KFA-1000[b]	1990	Panchromatic	7
			Multispectral	10
MIR-Priroda, Germany	MOMS-2P	1996	Panchromatic	6
Israel	EROS	Mid 1997	Panchromatic	1.5
	DAVID	1998	Multispectral?	5
NASDA, Japan	AVNIR[c]	1996	Panchromatic	8
	AVNIR-2	2002	Panchromatic	2.5
			Multispectral	10
CTA/NASA, USA	Clark	Early 1997	Panchromatic	3
TRW/NASA, USA	Lewis[c]	Early 1997	Panchromatic	5
			Hyperspectral	30
US Military, USA	CORONA, LANYARD[b]	1961–1972	Panchromatic	>1.8
NASA, USA	EOA AM-2 (Landsat 8)	2004	Multispectral	15

[a] Actually a series of latest is No. 7 satellites.
[b] Based on returnable cameras and film rather than electronics.
[c] Mission failed.

the internal structure of the plant responsible for reflecting the radiation, some transmission occurs but there is very little absorption. Radiation that is transmitted through the canopy of the plant will be reflected by any layers underneath, increasing the overall amount of reflected light. So we can see that the more internal structure a plant has the more radiation it will reflect and the more leaf layers there are the greater the reflection will be. In the middle-infrared water absorption is the controlling factor, and plant water content is a function of leaf thickness and percentage moisture in the leaf. It follows that as different species/families of plants have different structures they will have different reflectance characteristics across all parts of the electromagnetic spectrum but that some wavelengths provide more information than others. It also follows that the same species will exhibit different properties at different times of the year and in different growing conditions. For example, in early spring woodland (as the leaves are just emerging) it may be difficult to distinguish deciduous from autumn sown cereals; later on in the summer it may be possible to distinguish cereals from rape seed or sugar beet. As a more exotic example of fine attribute resolution, Ramsey and Jensen (1995) discuss distinguishing between three species of mangrove (red, white and black) and four types of forest (basin, overwash, fringe and riverine) in Florida.

The precise form of the reflectance across the spectrum that is produced by a particular target is called the *spectral response curve*, sometimes the **spectral signature**. The term signature implies uniqueness. Unfortunately this is rarely the case at the spectral resolution of most common sensors, so the term signature should be used with care. As described earlier, at the spatial resolution of many sensors the measured reflectance can be the result of the response of several disparate surfaces. *Mixture modelling* is the process of attempting to separate the response of several different surfaces to determine what mixture of surfaces is present within an individual pixel.

Having said that we can characterise how different types of vegetation reflect light of different wavelengths we then have to contend with the fact that it changes with time. As plants grow their characteristics change. In the early stages there is not much tissue to reflect light; by flowering the biomass will have increased considerably and there is lots of chlorophyll, internal structure, leaf layers and internal water. At senescence the moisture is gone and so is the chlorophyll and the internal structure has collapsed. Far from causing problems, these temporal changes improve the discrimination of vegetation types. For example, the ITE Land Cover Map of Great Britain produced from Landsat TM data makes use of winter and summer scenes to discriminate different land cover types. Summer and winter scenes are used primarily to distinguish between arable land and agricultural grasslands, but also have a role in distinguishing different types of forestry.

Similarly the relationship between environmental conditions and plant growth will be reflected (no pun intended) in their spectral response curves. This is exploited to predict plant yields and to monitor stress in vegetation. Much of the background research has come from ground-based radiometry, although the potential for wide-scale application has been demonstrated.

As well as vegetation the reflectance from soil may be of interest in ecological and environmental studies. Reflectance from soil is governed by a number of interacting variables such as texture, organic matter, mineral content, moisture and surface treatment. In general terms wet soils reflect less (especially in the infrared end of the spectrum): a clay soil has a lower reflectance than a sandy soil. A high organic matter content will lower the reflectance and so will iron oxides. A rough soil surface reflects less than a smooth surface (strictly speaking it does not reflect less, but it scatters the reflectance more, so that in some directions the signal is lower).

It is important to remember that the response within each pixel is not independent of the surrounding land cover. This means that within a patch that has a homogeneous *land use* the sensor may detect several different types of land cover. For example, a bare patch in an arable field caused by poor establishment of the crop or disease or pest attack may be characterised as being very different from the surrounding area. Of course this heterogeneous response may be very useful and the main reason why you are using remote sensing data; alternatively it could be a serious problem.

A further problem with remote sensing data is that topography affects the reflectance. Areas that are in shadow will obviously reflect much less than those in bright sun; in some cases it may be difficult to determine surface types because the signal is too weak.

Photographs are occasionally commissioned specifically for ecological research, but images from satellites can never be specifically commissioned for individual projects because of the horrendous costs involved. However, sensors such as the AVHRR sensor for vegetation biomass and the ocean colour sensor have been launched that appear to have no other function than environmental and ecological studies.

Satellite sensors

Most remote sensing satellites are in a Sun-synchronous near-polar orbit about the Earth. This means that they travel from north to south crossing the Equator at the same solar time each day, so the angular relationship between the Sun, satellite and centre of the field of view is more or less constant (it will vary between seasons). Landsat Thematic Mapper (TM) and the SPOT sensors orbit at an altitude of 700–800 km.

As the Earth is rotating at the same time as the satellite is orbiting, the **track** advances across the surface of the Earth. This means, in the case of Landsat TM, that subsequent paths are 2752 km apart and a 16-day period elapses before the satellite crosses over the same piece of ground again.

The area of ground covered in an image is known as the *swath*, the width of which varies according to the sensor (the swath is sometimes referred to as the *total field of view*). In the case of Landsat TM the swath is 185 km wide. Although the sensor is continuously imaging the Earth the data are divided up into approximately square chunks and referenced to by paths and rows.

Most sensors look straight down at the Earth, the nadir view. SPOT is also capable of imaging off-nadir. SPOT carries two sensors mounted side by side and can be programmed to point up to ±27°. This has two advantages: first, it is possible to look sideways and image areas that are adjacent to the path of the satellite; second, the two images can be used as a stereo pair for topographic mapping.

The sensing instruments on-board the satellite are various types of photoelectric devices; that is, they generate a current when photons hit them. Different wavelengths are detected using different materials and some sensors contain systems of mirrors, diffraction gratings and filters to direct the incoming radiation of different wavelengths to the appropriate detector. The signal generated by the incoming radiation is then translated into digital numbers or the pixel values that make up the image; typically values are recorded with eight bits, corresponding to 256 discrete values.

A *pushbroom* scanner is a solid-state sensor with no moving parts; it uses an array of charged-coupled devices, whereas TM uses an opto-mechanical device in which a rotating mirror directs the incident radiation onto the detectors. A satellite image is often treated as a raster product, which implies homogeneity within the cell; however, the sensor may well be biased so that, for example, features that are near the centre of the cell have more influence than features on the periphery.

Most ecological studies make use of data collected by sensors working in the visible and near-infrared parts of the spectrum, Kasischke *et al.* (1997) review the much more limited use of radar information in ecological surveys. The main uses for radar (particularly synthetic aperture radar, SAR) in ecological work appears to be estimating forest biomass, estimating the extent of flooding in vegetated marshes, identification of freeze–thaw cycles in high latitudes and identification of some types of land use change. Estimating forest biomass by radar appears to range from the very good to the pretty awful, depending on the type of forest (or the amount of research that has been carried out so far). Work is continuing in this topic; strangely (to us) it appears to work better with complex tropical forests than where a single species is dominant. Detecting flooding under a forest canopy appears to be a problem 'made' for radar; good accuracies can be achieved and every other method are time-consuming, expensive and error-prone. Because of the very different properties of ice and water, radar can detect the freeze–thaw cycles in high-latitude environments; changes may be important for all sorts of global environmental monitoring. Classification of land cover by radar is primarily based on surface roughness, vegetation structure and moisture conditions, whereas multispectral imagery relies on reflected solar illumination or surface temperature. The relative merits of the different systems are the subject of on-going research.

Types of error

Treatment of error and uncertainty is one of those areas of endeavour that is much talked and written about but in practice often ignored. You could argue that what separates the arts (and pseudo-sciences) from the sciences is whether the observations or experiments are repeatable. We know that physics is a science because given the necessary resources. We could repeat any physics experiment and repeat the observations. We know that history is not a science because cause and effect are subject to interpretation. Whether we view your ecological research as science or pseudo-science will depend on how rigorous your data collection was. In essence, sources of error can be reduced to three main types: accidental, systematic and random.

Accidental or gross errors can be as simple as writing down 153 instead of 135, or maybe it was the end of a long cold day and even you are not sure what your handwriting says. It is important to remember that accidental errors will always occur. With care and attention you can reduce their frequency, but the only way to eliminate them is to have independent measurements of the quantity in question. The (only) nice thing about accidental errors is that the larger they are the easier they are to detect.

Systematic errors generally arise from the equipment or technique being used. For example, you might be using an old tape measure that over the years has stretched so that every distance you measure is slightly too short. Alternatively you might be one of those people who always rounds numbers up (or down) instead of to the nearest odd number.

Random errors (sometimes called 'noise') occur at the limit of precision of the instrument or your observation technique. Random errors are the only type of error that is amenable to statistical analysis. Taking repeated measures of a variable and calculating and using a mean value implies that the errors are random. If your measurements include accidental or systematic errors the mean value will not necessarily represent a better estimate of the true value.

The general principle of reducing errors is that you need to take several independent measurements of any variable which is critical in itself or to other measurements. The Chinese are supposed to have a proverb: 'one son is no son, two sons are half a son, three sons are one son'. Such a proverb should be adapted for measurements: one reading tells you nothing (there could be accidental, systematic, random errors), two readings are half a measurement (all you know is that the two readings are different), three readings are a measurement (an accidental error stands out, random error is amenable to statistical analysis but beware: systematic errors are still 'untamed').

A second principle always to bear in mind in any data collection exercise is redundancy. If a location can be determined in several independent ways then each of the types of error (accidental, systematic and random) can be countered. If time and resources do not allow rigorous data collection then at the very least you must do some 'quality assurance' measurements.

Importing data into a GIS

Most GIS use a 'field' model of data storage (Chapter 2) using either raster or vector format. Object-oriented GIS are beyond the scope of this work and will not be discussed further. In essence spatial data consist of one or more attributes at a specific location. The location can be determined using either explicit spatial co-ordinates such as *x–y*, latitude–longitude or easting–northing, or implicitly such as row–column identifiers. The term attribute is used rather loosely in GIS and can apply to what is being represented (a tree, a river) or how it is being represented (a point, a line).

Each GIS typically uses its own data format, which is optimised in some way to reduce storage, allow efficient manipulation of data or avoid law suits (because someone else has patented the 'obvious way to do it'). However, it is usually possible to convert between systems without too much trouble, although it is often necessary to use a special export format to ensure that all the data are transferred.

There are many possible electronic data formats. Table 4.5 shows a few of the more common (and less common formats).

Table 4.5   Common (and not so common) electronic formats

Format	Description
ADS	Automated Digitising System US Bureau of Land Management – sub-system of MOSS
ADRG	Arc Digitised Raster Graphics
AMS	Automated Mapping System US Department of the Interior – sub-system of MOSS
BIL,BIP, BSQ	Bit Interleaved by Line, Bit Interleaved by Pixel, etc., ASCII data description files
dBase	database file
DIME	Dual Independent Map Encoding US Bureau of Census; replaced in 1990 by TIGER
DLG	Digital Line Graph US Geological Survey – digital version of USGS topographical quadrangle maps
DTED	Digital Terrain Elevation Data US Defense Mapping Agency format
DXF	Drawing Interchange Format Standard for CAD packages
E00	ESRI's ARC export format (for image processing) When importing you need to know if the coverages are lines, points or a grid
ERDAS	ERDAS proprietary format
ETAK	MapBase file Digital street map (USA only)
GIRAS	Land Use and Land Cover Data US Geological Survey format for land use and land cover

Table 4.5 (*continued*)

Format	Description
GRASS	Geographic Resource Analysis Support System
IGDS	Interactive Graphic Design Software Intergraph's Microstations format (also called design files)
IGES	Initial Graphics Exchange Standard US Department of Commerce format
IMAGINE	ERDAS proprietary format (for image processing)
ISIF	Intergraph Standard Interchange Format Originally a proprietary system but now more widely used
MIDAS	Map Information Assembly Display US Department of Agriculture Soil Conservation
MOSS	MOSS export file ASCII files readable by US Department of Interior – public domain GIS called MOSS
NTF	National Transfer File Ordnance Survey format
RLC	Run Length Encoded Usually for scanned monochrome images
SDTS	Spatial Data Transfer Standard US Federal Information Processing Standard 173
Shapefile	ESRI's ArcView format
SIF	Standard Interchange Format Primarily before editing and printing
SLF	Standard Linear Format US Defense Mapping Agency format
Sun Raster	Sun raster file Used on some UNIX workstations; data stored in 1, 8, 24 or 32 bits per pixel
TIFF	Tag Image File Format Desk-top publishing format
TIGER	Topographical Integrated Geographic Encoding and Referencing File US Bureau of Census
VPF	Vector Product Format Developed for the US Defense Mapping Agency as part of the digital chart of the world

Editing and cleaning data

The two most important issues to consider once you have imported the data into the GIS are whether they are topologically consistent and whether they are in the correct place.

One of the great advantages of raster data is that there is no topology to check for consistency. With vector data you need to confirm that polygons really are polygons, that lines meet or do not meet at the required place and that boundaries that should be in common are in common. When data have been collected as a number of tiles or sheets it will be necessary to 'edge match' where features cross the sheet edges. Speaking from bitter experience we would very strongly urge that if you are involved in a project that uses digitising to capture data, the person doing the digitising should be forced to do the topology checking. A very common mistake when checking for topological consistency is to use visual inspection to decide whether a polygon is really closed (is consistent). However, topology is in the 'mind' of the software; the only way to check is to request that the software identify what it thinks are polygons and then check that against what you think should be polygons. Often checking for topological consistency is carried out in a process called 'on-screen digitising'.

Whether you use raster or vector data or both you will need to check that they are in the 'correct' place. It is possible to use data related to an arbitrary co-ordinate system. When data are collected from a small site the effort needed to relate the co-ordinates to a 'standard' system may seem an unnecessary amount of work, but it is usually pays in the long run.

Scanning

Existing maps and plans on paper may be converted into electronic format by scanning. Most scanners were designed for 'artistic' rather than scientific work and are rather unintelligent. A simple scanner works rather like a photocopier, except instead of putting the results on a mechanical drum for immediate printing they are stored in the computer as a matrix of numbers. The values generated by the scanner are the values of the reflectance or colour of the image being scanned; for analysis it is necessary to convert these values into meaningful classes. Even when the scanner does not introduce geometrical distortions it will be necessary to rotate and scale the output so that it is spatially correct. It may also be necessary to edit the output manually, especially where the original document includes text. Also be aware that the map projection used on the original map may be different from the one used on your other data sets.

Digitising

The density of information on a printed map is prodigious. Digitising is one process of converting co-ordinates printed on a map into co-ordinates in the GIS without having to read the co-ordinates directly and type them in.

Box 4.6   Glossary of digitising terms

arc – a line (it does not have to be curved!) consisting of a series of **_vertices_** joined by straight sections

cursor – see **_puck_**

dangle – the most common error in digitising, where lines cross instead of meeting; needs to be corrected by **_snapping_** the **_nodes_** together

node – the end points of a line (special case of a **_vertex_**). If two lines cross but do not share a node they will not interact

puck – the pointing device on a digitising **_tablet_** (roughly equivalent to a mouse on a computer; also called a **_cursor_**). The cross hairs on the puck are lined up on the point to be captured; when a button on the puck is pressed the location on the tablet is recorded

tablet – a table that has many fine wires embedded in it for detecting the magnetic field generated by the **_puck_** (and so locating the position)

snapping – process of correcting digitising errors by making lines or points that should join 'snap' together; it may be necessary to insert another node first

vertex – a point on a line with known co-ordinates

Digitising has its own terminology, Box 4.6 provides a quick glossary of some of the terms you may encounter.

Manual digitising

Digitising is a method of converting all or part of existing maps and plans into electronic form. A digitiser consists of a special tablet, a pointing device or **cursor** and software to store and display the results. Embedded in the tablet are a series of very fine copper wires crossing at right angles to each other; when a button on the pointing device is pressed it generates a magnetic field so that the location of the pointer can be determined to within fractions of a millimetre. Before digitising the contents of the map it is necessary to select points on the map with known co-ordinates to allow the transformation between the table co-ordinates and the map co-ordinates. It is usual to select six points near the corners and the middle of the longest sides, although three points are sufficient geometrically. The operator is responsible for tracing around each feature of interest, pressing one or other of the buttons on the cursor each time the line changes direction. Digitised software varies considerably in how much information the user supplies during the digitising process. On some systems the user specifies only whether the feature is a line or an isolated point and supplies an identification number; with such systems the attributes of the feature are assigned in a later operation. Other systems enable or require the user to supply much more information about the feature as it is being digitised. The length of time it takes to digitise a map is obviously dependent on the complexity and number of features to be collected and the experience and skill of the operator but it always seems to take longer than expected! With an inexperienced or careless operator it can take longer to make the results topologically correct and attach the attributes than to digitise the map.

What can go wrong with digitising

Digitising is rather like knitting: you need to plan what you are going to do first, carry out a test piece to make sure you understand what you have to do, carry out the task consistently and carefully and never count up how many 'stitches' more there are before you have finished. So what can go wrong? Problems derive mostly from poor concentration and poor preparation. Box 4.7 provides our advice on 'painless digitising'. Figure 4.4 illustrates some of the things that you will have to sort out at the end of each day.

Box 4.7   Advice on painless digitising

1. Decide on what you *need* to digitise and in what order you are going to do the work. When a feature has a natural direction, such as a river, decide if you are *always* going to work upstream or downstream.
2. Make sure your map can be laid flat. If it has been out in the field and has crumpled and folded try running a moderately hot iron over it, if it still will not flatten think about cutting it up into smaller sheets that will go flat.
3. Fix your map very firmly onto the tablet with masking tape (not plastic tape, double sided tape, chewing gum or worse).
4. Tilt the tablet to a comfortable angle: your line of sight when you are working *must* be perpendicular to the digitising tablet so that there is no possibility of parallax error.
5. Digitise at least six control points (the four corners and the middle of the two longest edges); if the map is not absolutely flat or if it is a photocopy add a few more control points. Note that while three points are geometrically sufficient (provided they do not lie in a straight line) good practice is to use six. Check that the root mean error on the control points is acceptable; the accuracy of everything else you are digitising depends on these control points.
6. Start work – mark off the digitised features with a highlighting pen or felt tip pen as you go along (this helps make sure you have not missed anything or done something twice, and it reminds you that you *are* making progress).
7. Where a curve and a straight section meet it can be difficult to decide exactly where the transition occurs; play safe and bracket the probable location, be liberal with the number of points on a curve (you can always weed some out, but having to add extra is no fun). When the line is very thick digitise the centre of the line.
8. Try and establish a consistent rhythm to working; some people like to play music, but we find it distracting (must be showing our age).
9. Stop every hour for a 5 min break to rest your eyes and stretch your muscles.
10. Import the data into the GIS and remove dangling nodes, sliver polygons and other 'funnies' at the end of each day before you go home. *Insist* that the person doing the digitising is the person who sorts out any problems with the data.

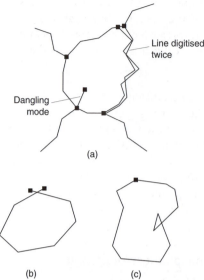

Figure 4.4 Digitised features that will need cleaning: (a) line digitised twice and dangling node, (b) polygon not closed, and (c) weird polygon.

Semi-automated digitising

It is a fairly common belief that digitising can be automated; unfortunately this is not the case. Semi-automatic methods usually rely on scanning the map to produce a high-resolution raster image (at 400 dots per inch or more). A line-following algorithm then searches the image and extracts lines. Such methods should be termed semi-automated because algorithms cannot interpret topology and even the best 'gap-jumping' and junction-ignoring algorithms are far from foolproof. Most large national mapping organisations have a semi-automated digitising capability. The cost of the hardware and software are beyond virtually everyone else.

Commercial digitising

Because of the cost of even a small digitising tablet and its specialised nature it may be cost-effective to employ one of the commercial companies that offer a digitising service.

Image processing

Remotely sensed data captured on photographs or electronically must be processed before they can be used. Image processing is a specialised task that you are unlikely to become involved in (unless you are particularly interested), but it is probably useful to have some idea of what is involved in the process. Four stages are usually needed:

- contrast stretching,
- filtering,
- geometric correction, and
- classification.

Contrast stretching

The big difference between film and electronic images is that with film the exposure time is manipulated as the data are collected so that as large a range of values as possible are recorded. Under- and over-exposure restricts the range to an upper and lower range of possible values, although some 'stretching' is possible by manipulating how the film is developed. Most electronic sensors record data in a number of spectral windows or bands, each one supplied separately. Typically values in a band can only take values between 0 and 255, but in its raw form values may occupy only a narrow range of values. The simplest contrast stretch is a linear stretch based on the upper and lower bounds:

new_value = (old_value − old_min) × 255 / (old_max − old_min)

More complex and robust stretches exist that are less influenced by a few extreme values.

Filtering

Having stretched the image so that the full range of values is used it is usually necessary to filter the image. Filters are used for a variety of purposes to improve the quality of the image. Where there is a lot of 'noise' in the image a *low-pass* filter is used to reduce local extremes. Where it is important to identify boundaries between patches a *high-pass* filter is used to increase the contrast locally. A filter consists (conceptually) of a small matrix (usually three by three cells) of 'weights'. This is moved over the image one cell at a time and the new value for the centre point is calculated as the mean of the weights multiplied by the cell values. Adaptive filters replace the value of a cell only if some criteria are met. A typical use might be to remove (or identify) 'odd values' as being more than some multiple of the standard deviation away from the local mean value. An adaptive filter may use local (neighbourhood) values or global values in the tests. Box 4.8 shows a number of commonly used filters. Filters have other uses than improving image quality; for example, they are frequently used on digital terrain models (DTM) before hydrological functions are applied.

Geometric correction

Geometric correction of an image can be visualised as the process of rotating and scaling an image so that the original geometry of how the image was captured is recreated. Remote sensing from a satellite is relatively easy to adjust geometrically; however, similar instruments mounted on an aeroplane make

Box 4.8   Some common image analysis filters

Low-pass filter to smooth out 'noise' and small imperfections in an image, also used on digital terrain models before performing estimating hydrological functions (see Chapter 5).

$$
\begin{matrix}
1 & 1 & 1 \\
1 & 1 & 1 \\
1 & 1 & 1
\end{matrix}
$$

High-pass filter (this is a *Laplacian* filter) to 'sharpen' an image to emphasise the edges between patches.

$$
\begin{matrix}
0 & -1 & 0 \\
-1 & 5 & -1 \\
0 & -1 & 0
\end{matrix}
$$

Non-linear Sobel filter, a more sophisticated version of the Laplacian filter.

$$
\sqrt{\left[\begin{pmatrix} -1 & 0 & 1 \\ -2 & 0 & 2 \\ -1 & 0 & 1 \end{pmatrix}^2 + \begin{pmatrix} 1 & 2 & 1 \\ 0 & 0 & 0 \\ -1 & -2 & -1 \end{pmatrix}^2 \right]}
$$

Adaptive filters:

If $[(v - \mu)^2 > F\sigma^2, \mu, v]$

that example could be used to remove 'outliers' in the data: if the value of the cell, v, is much bigger (or smaller) than might be expected then replace it with the mean, μ, otherwise leave it alone. In this case the test of what is expected is whether the deviation from the mean is more than some factor, F, multiplied by the variance, σ^2.

geometric corrections horrendously difficult. An image from an aeroplane-mounted scanner requires that every pixel be corrected for the yaw, pitch and roll rotations of the aircraft as well as its movement forward, up and down and sideways and the relative elevation of the imaged point. For an air photograph the whole image is geometrically coherent so the positional unknowns can be deduced from the relative position of known *ground control points*. Because a satellite is moving through what amounts to a vacuum the scanned image is, like a photograph, geometrically coherent. Note that over the swath width of a typical satellite image (185 km for Landsat TM) the curvature of the Earth and atmospheric refraction both need to be carried out (usually done by the suppliers of the imagery).

Geometric correction of a coherent image is done by rotating and scaling the image so that points on the image coincide with the ground control points. Ground control points must be well distributed across the image especially near the corners and edges. Ground control points can sometimes be put out before the imagery is collected for air photographs, but given the resolution of most satellite imagery identifiable targets need to be very large. Identifying good ground control points after the imagery has been collected has the advantage that they are in good locations. Identifying points on the ground can be surprisingly difficult even in this country. From personal experience one of the authors can vouch for how difficult it is to identify ground control points from air photographs in some locations; attempts to identify particular boulders in the deserts of northern Sudan, for example, have been known to provoke all sorts of profane language!

The success of the geometric correction is assessed by checking the *root mean square* (RMS) value. Note that unless you have redundancy in the number of ground control points that any misidentified point will be difficult to identify and rectify. After the image has been geometrically corrected the pixels will not necessarily be square and the image needs to be re-sampled. Several different interpolation techniques can be employed: *nearest neighbour* techniques maintain the original classes exactly by allocating single values, while *bilinear* and *cubic convolutions* estimate some sort of mean value. Interpolation techniques are discussed in more detail in Chapter 5.

Classification

Having completed the contrast stretch and geometric corrections you will be ready for the final stage of processing: classifying the image to try and relate reflected light to ecological phenomena (vegetation type, biomass and so on). Two principal methods of classification are used, supervised and unsupervised. Supervised classification requires the user to specify what classes are present and identify homogeneous training patches of each class. The training classes are used within the classification process to allocate the remaining pixels to one of the classes. With unsupervised classifications, classes are derived from the data and then the user has to determine what the classes mean. Classification algorithms are covered in Chapter 5.

Aggregating data

It is sometimes necessary to use data that have been aggregated. Aggregating data can introduce some 'interesting' discrepancies; below we conduct a 'case study' to illustrate some of these problems.

On 'GIS Island' we have a very simple ecology and a very simple geography. Figure 4.5 illustrates the number of species of water plants in each county (all of which happen to be rectangular). Notice that there is a very simple trend of increasing diversity from north to south and there are three biodiversity hot spots: one in the north-west, one near the centre and one in the south-east. You might also notice that each hot spot has exactly three times the average number of species in the counties surrounding it (we said it was a very simple island).

Suppose that we suspect that there is some relationship between the diversity of water plants and the diversity of dragonflies. Figure 4.6 is a map of dragonfly diversity. It has biodiversity hot spots in the same place as the water plants and it also has increasing diversity from north to south. On 'GIS Island'

8	2	2	2	2	2	2	2
3	3	3	3	3	3	3	3
4	4	4	12	4	4	4	4
5	5	5	5	5	5	5	13

Figure 4.5   GIS Island: number of water plant species per county.

4	1	1	1	1	1	1	1
3	3	3	3	3	3	3	3
4	4	4	12	4	4	4	4
10	10	10	10	10	10	10	26

Figure 4.6   GIS Island: number of dragonfly species per county.

dragonflies are more susceptible to the cold than plants so that in the north there are exactly half the number of dragonfly species as there are water plant species; in the south there are twice as many dragonfly species as plant species, while in the 'mid-latitudes' there are equal numbers of dragonfly species as plant species.

Suppose that instead of the 32 counties we require a simplified (aggregated) map of four regions for the four GIS island members of parliament, or the four conservation groups, or because there is a strict page number limit in the journal or conference you want to submit your work to. Obviously, you say, there must be a simple north–south trend in the aggregated data – not necessarily! Below are three different aggregations. If our four regions are aligned north–south then as you would hope you get a north–south trend, if we aggregate into north-east, north-west, south-east, south-west then there is a clear trend from a low ratio in the north-west to a high diversity in the south-east. However, if we aggregate in four regions east–west the trend of increasing diversity runs from west to east.

Aggregation 1 Ratio of dragonfly diversity to plant diversity by 'region'

0.5
1.0
1.0
2.0

Aggregation 2 Ratio of dragonfly diversity to plant diversity by 'region'

0.54	0.76
1.64	2.32

Aggregation 3 Ratio of dragonfly diversity to plant diversity by 'region'

1.15	1.22	1.29	1.44

GIS Island is of course an artificial data set designed to illustrate the point that the method of aggregating the data affects the pattern of trends, but almost all published data are aggregated in some way. Do you know why it was aggregated in that manner? Do you know what the trends are? If our four regions had been irregular in size and shape would you have immediately recognised the artificiality of the situation?

Geo–correction

The whole point about a GIS is that the data are spatially coherent. After the data have been collected they must be geo-corrected or fitted to a known co-ordinate system. The process of geometric correction is sometimes termed *rubber sheeting* or *warping*. Warping comes from the analogy of drawing the data on a rubber sheet and then pulling the edges of the sheet this way and that until the image conforms to a given pattern. Rubber sheeting algorithms exist in most GIS but should only be used as a last (desperate) resort. About the only time they should be used is to take out the distortions due to photocopying. The processes needed to geo-correct data collected from a small area (less than a few tens of kilometres) are shift (translate), rotate and scale; if the data cover more than a few tens of kilometres it will also be necessary to apply a projection as well.

Consider the case where you have collected some data in the field using field survey methods to measure distance and angles from some point to the features of interest. Suppose that you have then digitised the data using an arbitrary co-ordinate system. The first process is to locate your origin in the desired co-ordinate system (most simply by taking observations from known points). Then 'shift' all your data so that your arbitrary origin coincides with its location in the desired co-ordinate system. Second, you need to relate your arbitrary zero direction with grid 'north' in the desired system and rotate all your observations about the origin so that the two directions coincide. Finally, you need to scale your observations so that distances between points represent distances on the 'mapping plane', not the actual (real) distance measured. This process is illustrated in Figure 4.7.

If the data cover more than a few tens of kilometres then it will be necessary to apply the appropriate projection. To apply a projection you need to

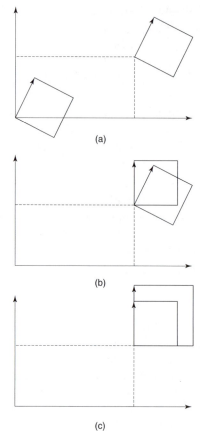

(a)

(b)

(c)

Figure 4.7 Geo-correction of data from a small area: (a) translation/shift, (b) rotation and (c) scaling.

know three things: which projection is being used (Transverse Mercator, Alber Polyconic, State Plane, etc.), which set of parameters is being used for the projection (where the 'standard parallel(s)' are, where the origin is, which scaling factors are being used, etc.), and which spheroid is to be used (WGS84, Clarke, Everest, etc.). The actual mathematics of projections is more than a little complicated; fortunately, most GIS include at least the most common systems.

Review

A GIS is nothing without good data. In this chapter we attempt to introduce you to some of the common and not so common sources of existing data. You need to be aware that other people's data will almost always have been processed at least once, sometimes several times. Even if you have the luxury of collecting your own data you still need to be aware of what data were collected, why and how.

Further reading

If you are really interested in the 272 regulations relating to copyright in the UK in 1997 then the best starting point may be M.F. Flint *A User's Guide to Copyright* (Butterworth, London, 1997) and good luck to you! There is a large literature on remote sensing and several active journals; we try and keep an eye on the following: *International Journal of Remote Sensing, Photogrammetric Engineering* and *Remote Sensing,* and *Remote Sensing the Environment.* It is difficult to guess how long a satellite sensor will keep working; some fail after hours or days, others continue working for years longer than was expected. The Internet is usually a good source of information (and example images) from different sensors. A reasonable introduction to the Internet is J.R. Levine, C. Baroudi and M.L. Young *The Internet for Dummies* (IDG Books, Foster City, Calif., 1996).

## Chapter 5

Getting to grips with GIS

Getting started

The critical problem you are about to confront is the large number of commands that are included in most GIS. A workstation package might need 2–3 m of shelving for the printed documentation and even a PC package will contain dozens of commands. Perhaps the first thing to say is: '*don't panic*', you only actually *need* 4–5% of the commands (although different people tend to use different sub-sets of the total).

We have tried to illustrate getting to grips with GIS with specific examples from three contrasting systems:

- ArcInfo® Version 7 for Unix workstations. Version 7 is arguably the largest and most comprehensive GIS currently available. It is widely used in academia and industry (and has one of the most active newsgroups on the Internet). ArcInfo has its own well-developed macro language AML. The Environmental Systems Research Institute (ESRI) Inc. offers substantial educational discounts; more details of its products (and example applications) can be found on the Internet – **http://www.esri.com**
- GRASS Version 4 for Unix workstations. Developed by the US Military's Construction Engineering Research Laboratories (CERL). The Geographic Resources Analysis Support System is free! and you are provided with *all* the source code (which is written in C). It has very strong raster processing capabilities and use can be made of the standard Unix shell scripts for constructing macros. Further details can be found on **http://www.cecer.army.mil/grass**
- Idrisi for Windows Version 1. A compact PC-based system; it has very good raster and image processing capabilities. It has a very 'clean' menu interface but a rather basic macro language. Developed by Clark University, with help from the United Nations; substantial education discounts are available. Further details are available on **http://www.idrisi.clarku.edu**

Independently generated examples, information, work books, lecture notes and slides are provided for all three systems by Project Assist (Academic Support for Spatial Information Systems) based in the Department of Geography at Leicester University (Leicester LE1 7RH, UK, or **http://www.geog.le.ac.uk/assist/index.html**).

The two common ways of interacting with GIS software are by command line and menu systems. Menu systems (and graphical user interfaces) offer

Box 5.1 Make life easy for yourself!

Your GIS should allow you to use *shell scripts, batch files* or *macros*. These are all slightly different terms for the process of storing a string of commands in a plain text file; the script can then be executed (run) when it is convenient. We *strongly* advise that you start to use scripts as soon as you start to use a GIS. Scripts offer many advantages:

1. Help you remember how to do something, what you have done and what you have yet to do.
2. Help you explain what you have done and provides some quality assurance (especially which version of the data you used, what types of process were applied and the order that the process took place in).
3. Save a great deal of typing and retyping (especially for commands where you have to set several flags or arguments).
4. Free you to think about the problem while the computer gets on with the number crunching.
5. Can be reused or edited on future projects months or years later, so you can build up your own library of useful tools.
6. Provide the basis for documenting your project (one of those important tasks that the longer you leave it, the worse it gets).

Simple rules for shell scripts:

- always start with a comment on what the scripts are supposed to do,
- use blank lines, lower case text and indented lines to make the scripts easily readable, and
- if they get longer than about two pages break them down into smaller pieces.

Examples of scripts are included in the last section of this chapter.

---

some advantages: they look impressive, they reduce the amount of typing you have to do and they provide plenty of 'clues' and reminders as to what is possible (and what the commands are called!). Provided that the menu system allows you to record a sequence of commands that can be rerun at a later date (after you have added your comments) they should not do too much harm. However, most proprietary menu systems are like stabilisers on a bicycle – sooner or later you need to learn how to do it properly. Initially the command line method of interacting with the software is more cumbersome than using a menu system, and the advantage of command line interaction is apparent only when it is used in shell scripts (batch files, macros – see Box 5.1). If you do not use shell scripts then command line interaction is a pain in the proverbial. Commands may have optional arguments or flags in addition to the mandatory ones. In GIS like ArcInfo® and Idrisi the order of the arguments is critical; items that are not required are replaced by a 'place holder' number (in Idrisi the 'place holder' is usually a zero). In a GIS like GRASS the order is not important but each argument must be prefixed by the correct name and an equals sign. Table 5.1 provides a few examples of the different styles used.

Table 5.1   Examples of different command line styles

Activity	GIS	Command line argument
Define the geographic extent of interest by co-ordinates[a]	GRASS	g.region w=0 e=100 s=0 n=100
	ArcInfo	mapextent 0 0 100 100
	Idrisi[e]	initial × friction 2 2 1.0 2 100 100 plane m 0 100 0 100 1 m
Draw a raster grid on the screen[b]	GRASS	d.rast rast=map1
	ArcInfo	gridshade map1 (alternatively image map1)
	Idrisi[e]	display × a map1 Idrisi256 y × 0
Put a buffer one cell thick around cells[c]	GRASS	r.buffer input=map1 output=map2 distance=1
	ArcInfo	map2 = expand (map1, 1, list, 1)
	Idrisi[e]	costgrow × map1 friction 1 map2 2
Perform an 'inverse distance weighting' interpolation using the six nearest points[d]	GRASS	r.surf.idw input=map1 output=map2 npoints=6
	ArcInfo	
	Idrisi[e]	map2 = IDW(map1, #, #, #, sample, 6) interpol × 2 sample_pts map2 m 1 2.0 Y 2 100 100

[a] The Idrisi command is also generating a new map (in this case called friction) of real numbers in binary format with initial values of 1.0 in all cells.
[b] With ArcInfo and GRASS you need to define a graphical canvas first (display 9999 1 in ArcInfo and d.mon start=×0 in GRASS). The Idrisi command also specifies the colour palette to use (Idrisi256) and that the title, legend and scale bar should be drawn (the y in fifth place).
[c] The Idrisi command requires a 'friction surface' (which in this case could be the map or image created by the initial command).
[d] The Idrisi command requires a 'vector' file of heights (sample_pts).
[e] In Idrisi arguments are preceded by an ×.

Because we actually follow our own advice and always, always use shell scripts or macros we could review the commands we have used over the last three years. We have used ArcInfo on a range of ecological and environmental projects, including:

• the impact of the electricity supply industry on Sites of Special Scientific Interest,
• the optimum (economic and environmental) route for a pipeline across a landscape,
• the effect of diatoms on sediment stability in the inter-tidal zone,
• the spread of alien weeds down river corridors, and
• the impact of climate change on agriculture.

In that time we have amassed a library of 358 scripts totalling 27 943 lines of AML code (in addition we also produced several menu systems and some awk, C and Fortran programs, but that is another story). Of the ArcInfo scripts:

• 8432 lines were comments, notes and descriptions;
• 6505 lines were blank lines inserted to improve the readability of the code;
• 7156 lines provided control for the flow of operations (logic and repetition);
• 2058 lines were concerned with loading data into and out of the relational database;

- 1731 lines were getting data into and out of ASCII data files and to and from the user;
- 1266 lines were cartographic commands for drawing, plotting and producing hard copy;
- 426 lines were conversion between formats and miscellaneous 'housekeeping' activities; and
- 369 lines were for spatial analysis and manipulation!

What the above analysis does not show (because we do not know!) is how many times a particular command was used in a script (because of all the conditional loops and 'if' statements); nor can we say how often we used a particular script. What you can see is that most commands issued are concerned with data management and with visualising, printing and plotting data rather than with spatial analysis. The commands that make a GIS unique are those concerned with spatial analysis and there are not many of them and they are mostly simple and easy to learn.

Of course one way to understand your GIS would be to read the manual. In practice we have to get pretty desperate to start on the manuals; in a way this is a pity because there are many things we do not know about because we have never even looked at some of the volumes in the manual. On the other hand neither of us has read the manual for the word processor used to write this book or for the computer or for the operating system for the computer. As long as you *always* make sure you have all your data *backed up* (and the backups backed up) then we suggest you start playing and look at the manual (or on-line help) to find out how to do something, not to learn what the software can do.

Classification of spatial analysis tools

Albrecht (1996) has devised a scheme for classifying spatial operators in a GIS, which is based on the user's view of how functions are related rather than on a technician's or software engineer's view. Under his scheme all spatial analysis can be described by 20 'classes' of operators; these can be combined to make even the most exotic applications (see Table 5.2). It should be noted that each of these operators exists in several different 'flavours'; for example, there are at least half a dozen ways to interpolate point data.

Albrecht's typology has much to recommend it, not least its user-centred view. An even simpler classification may be possible. Under the simplified scheme proposed here spatial analysis can be split into four functions:

Section A: measuring,
Section B: searching,
Section C: classifying, and
Section D: modelling.

Any classification is devised to be applicable or used for a particular set of circumstances or purposes. The classification proposed here is designed to help explain what a GIS can do for you and to provide some structure for a

Table 5.2 Albrecht's fundamental spatial operators

Search	1. Interpolation
	2. Search-by-region
	3. Search-by-attribute
	4. (Re)classification
Location analysis	5. Buffer
	6. Corridor
	7. Overlay
	8. Voroni/Thiessen
Terrain analysis	9. Slope/aspect
	10. Catchment/basin
	11. Drainage/network
	12. Viewshed
Distribution/neighbourhood	13. Cost/diffusion/spread
	14. Proximity
	15. Nearest-neighbour
Spatial analysis	16. Multivariate analysis
	17. Pattern/dispersion
	18. Centrality/connectedness
	19. Shape
Measurement	20. Measurement (geometry, statistics, topology)

description of spatial operators. The rest of this chapter is divided into these four major sections; under each section we describe some of the facilities we think (based on our experience) are likely to be of interest in an ecological study:

Measuring:	Individual features	Section A.1
	Terrain and other surfaces	Section A.2
	Pattern	Section A.3
Searching:	By region	Section B.1
	By attribute	Section B.2
	Interpolation	Section B.3
Classifying:	Overlay (raster & vector)	Section C.1
	Classification	Section C.2
	Statistics	Section C.3
Modelling:	Network	Section D.1
	Neighbourhood modelling	Section D.2
	Statistical modelling and Baysian inference	Section D.3
	Cost–diffusion–spread	Section D.4
	Map algebra	Section D.5

Section A: measuring spatial data

Quantitative measurement is the heart of GIS. As described in earlier chapters data are usually stored in either vector format (points, lines and polygons) or

Table 5.3 Typical measurement questions

Typical questions	Sub-class of operation	Section
How much land does Monks Wood Nature Reserve cover in total? How many open glades exist in Monks Wood and what area do they occupy? How long is the boundary between the edge of the forest and the fields under permanent pasture?	Measurement of individual features	A.1
What is the slope and aspect of each of the open areas in Monks Wood? Where is water likely to accumulate when it rains? Is it possible to see Wood Walton Fen from the edge of Monks Wood?	Measurement of terrain and other surfaces	A.2
What is the average distance between glades in the wood? Is the distribution of glades in the wood spatially random? How much area is available for an individual plant (how much territory can it occupy)?	Measurement of pattern	A.3

in raster format (grids, meshes or lattices). Whichever form of data storage is used, the measured values you will get are dependent on the resolution (precision) with which the data were collected, stored and manipulated. This makes estimating the precision and accuracy of the measurements you make a non-trivial exercise. Be especially careful when comparing areas or shapes of patches captured from data collected at different times by different people for different purposes. Be very careful of data that may have passed through a cartographer at some stage. Table 5.3 provides some 'typical' measuring questions.

Section A.1: measuring individual features

One of the simplest but most powerful things about GIS is the ease with which 'book-keeping' exercises can be performed.

Measurement of individual features is a matter of direct or indirect inspection of the underlying data. Direct inspection involves using a mouse to point at the object on the screen and using a command to return the value on one (or more) items at that point. Alternatively, the command may return the co-ordinates of the point of interest, which can be used subsequently to interrogate other data layers. For example, to return the value in a particular cell on a grid using the mouse to point to the map:

```
in ArcInfo: cellvalue map1 *
in Idrisi:   select the '?' icon from the tool bar
in GRASS: d.what.rast -1 map = map1
```

Similarly to measure the length of a line on a map:

in ArcInfo: `measure length *`
in Idrisi: `distance` (or `spdist` for spherical distances in radians)
in GRASS: `d.measure`

Indirect inspection of the database to identify the size or shape of a particular feature may involve listing or searching the underlying database table (ArcInfo) of 'header' files (GRASS and Idrisi). For example, in ArcInfo three commands that we use most frequently are:

- `describe map1` – lists type of feature, geographic extent (and co-ordinate projection system if any); for grid data lists the range, the maximum, minimum, mean and standard deviation of the data; for vector data lists the number of arcs, nodes, polygons, labels, etc.
- `items map1.pat` – (pat = polygon or point attribute table, integer grids have a `vat`) lists the items (columns) present in the data table (and how wide each column is) and whether the column holds integer, floating point or character information.
- `list map1.pat` – for each point or polygon lists all the information in the data table (the pat). For an integer grid `list map1.vat` provides the frequency of each 'class'.

For a large data set, listing a full data table and then trying to identify the feature of interest by eye is tedious, inefficient and error-prone. Although it might sound long-winded a better approach is to:

- display the data,
- identify features of interest by interrogating the database,
- redisplay the data to show which features have been selected (checking for errors),
- select one or more features, and
- interrogate the database to obtain the data.

For example, suppose we wanted to answer the question in Table 5.3 '*How many open glades exist in Monks Wood and what area do they occupy?*' from a vegetation map called map1 in polygon form. Assuming that the *arcplot* module has been started and that suitable cartographic commands have been set (map-extent, shade-set, line-type, page-size, page extent) and that 'glades' have a code of, say, 10 in the cover_type column of the data table.

- `Polygonshade map1 polys cov.rmt`
  {draw the land cover map1 with `cov.rmt` 'maps' classes to colours}
- `reselect map1 polys cover_type = 10`
  {select the polygons where the '`cover_type`' is appropriate;
  reselect will report how many polygons satisfy the query},
- `polygonshade map1 4`
  {draw only those polygons identified by the reselect command with whichever colour 4 happens to be so you can check that the selected 'set' looks reasonable}.

Having checked the right data have been selected exit arcplot and start the interactive version of reselect:

```
reselect map1 junk polys
  >: rsel cover_type eq 10
  >: <return>
  >: n
  >: n
list junk.pat
```

When attempting to measure an 'object' on a raster data set, an additional complexity arises through the necessity to aggregate (clump, group) adjacent cells to represent the feature of interest (in the ArcInfo regiongroup or the various zonal functions). Aggregation to produce clumps usually works well for features that naturally occur in large compact patches such as forests, wetlands and agricultural areas. Aggregation to identify linear features such as rivers or hedges are more difficult and it is usually only successful when the size of the cell is small compared with the minimum width of the feature.

Section A.2: measurement of terrain and other surfaces

GIS usually contain a specific sub-group of functions to measure the characteristics of surfaces. Many of the routines were developed for terrain analysis and hydrological studies but could be applied to any surface. We have split measurements of terrain into:

- measures of slope, aspect and curvature;
- defining catchments (basins) in the surface;
- defining drainage lines; and
- determining inter-visibility.

Slope, aspect and curvature of a surface

Within a GIS, surfaces can be easily manipulated as either grids (lattices) or by triangular irregular networks (TIN). Grids used specifically to represent topographic surfaces are termed digital terrain models (DTM) or digital elevation models (DEM). It should be noted that slopes are generally much less steep than you might imagine: a 10° slope is very steep, a flight of stairs is less than 45° and the recommended slope on a ladder only 76°. Because of the dependence of measures of slope on the resolution of the DTM the value of the 'mean slope' is difficult to interpret.

Quantities such as distance or temperature are measured on a linear scale, so such measurements can be summarised by statistics such as the mean. Quantities such as aspect are measured on a circular scale and *must* only be manipulated using circular statistics. For example, suppose you observe the direction that a bat leaves its roost each night for a week; on four nights it leaves in a north-easterly direction (bearing 45°) and on three nights in a north-westerly direction (bearing 315°). So the arithmetic mean direction is:

$$(4 \times 45° + 3 \times 315°) / 7 = 175°$$

but 175° is almost exactly due south and that does not make any sense given the observations.

Slope and aspect may be valuable to help predict the distribution of animals and especially plants, because of the way solar energy will vary (in hilly
regions precipitation is also likely to vary significantly with aspect). Concave
hill slopes will tend to concentrate surface water runoff, which could be
important for deciding where erosion might occur or where bogs and mires
might form. As well as the analytic function, measures of slope and aspect can
be a very powerful help in visualising the shape of a surface by modifying the
saturation or **hue** of the colour.

Defining catchments (and basins)

A catchment or basin defines an area that concentrates the movement of some
phenomenon to a single point, but the term is most often encountered in
connection with water. A surface water catchment defines an area of land with
the property that all excess water (from rain) flows out through a single point.

On the face of it calculating a catchment from surface data is simple (after
all you can sketch the boundary on a map in a few minutes). In practice finding catchments from digital data is difficult and good algorithms are complex.
It will be necessary to carry out some smoothing of the surface to remove
isolated (insignificant) peaks and hollows (pits). It is then usual to calculate
aspect and curvature and combine these values with the elevation to determine which cells drain into which. Because catchments are nested within each
other down to the limit of the resolution of the DTM even sophisticated
algorithms are very sensitive to the selection of the initial point. Although
typically applied to DTM and real elevations such an algorithm applied to an
accumulated-cost surface would help define patterns of influence.

Drainage lines of drainage network

A catchment defines an area that concentrates a phenomenon through a single
point. The term is most often encountered in hydrology but can apply to
other phenomena. Within a catchment, flow is not (in general) uniform but
will tend to be concentrated along certain lines (in hydrology, streams and
rivers). Assuming that the 'elevation data' exist in a raster format the 'direction of flow' can be estimated (in ArcInfo flowdirection) from the relative height of adjacent cells (some algorithms allocate all flow to the lowest of
the eight adjacent cells, other algorithms apply more complex schemes). By
estimating the cumulative through the 'flow direction' grid flow (in ArcInfo
flowaccumulation) a network of preferential flow can be estimated. By
selecting an arbitrary threshold value and then selecting those cells that exceeded
it [if (flow_acc > threshold) net = 1], a process that is illustrated
in Figure 5.1. Depending on the resolution of the 'elevation' data it may be
necessary to manipulate the selected cells further to ensure a complete network; most practical elevation data sets contain numerous local minima (pits)
that prevent a complete network being formed. The problem is that by smoothing out the local minima the network is distorted.

In ArcInfo given the elevation data in a grid called 'elev':

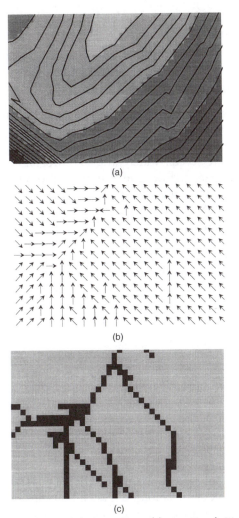

Figure 5.1 Flow accumulation and drainage lines: (a) elevation (with contours), (b) slope direction (averaged over 2 × 2 cells), and (c) accumulated flow (black cells show drainage area).

```
flow_dir = flowdirection (elev)
flow_acc = flowaccumulation (flow_dir)
{can concatenate these two steps}
if (flow_acc gt %threshold%) network = 1
{need to try a few values of threshold until the result is dense enough}
```

Inter-visibility and viewshed

Deciding whether two points are inter-visible depends principally on the presence of any obstructions and the elevation of the ground in between (in extreme cases it can also depend on the curvature of the Earth and the quality of the air). Calculating inter-visibility is a very intensive calculation on a computer. A viewshed is the area of land that can be seen from a particular point. Note that how visible an object will be is greatly dependent on whether or not it is on the 'skyline' and its colour.

Section A.3: measurement of patterns

There has been considerable interest in the field of landscape ecology in measures of landscape pattern, such as connectivity and fragmentation, but most GIS have yet to incorporate facilities to calculate these 'landscape metrics'. Fortunately, there are several widely available programs to calculate a wide range of landscape characteristics and that take various GIS types of data file. At least one of these programs, FRAGSTATS, can be downloaded from the Internet. FRAGSTATS provides 11 metrics of patches, 34 metrics of 'classes' and 46 metrics of landscape, primarily concerned with size, shape, edge type, core area and so on. So many possible metrics are a cause for concern, given enough measures one is bound (through random chance) to have a statistical (but not a causal) relationship with the organism being studied. We also worry about how the many organisms being studied can 'sense' some of these properties. This concern is forcefully put in the preface to the FRAGSTAT documentation (McGarigal and Marks, 1994):

> As the authors of FRAGSTATS we are VERY concerned about the potential for misuse of this program. Like most tools FRAGSTATS is only as 'good' as the user. FRAGSTATS crunches out a lot of numbers about the input landscape. These numbers can easily become 'golden' in the hands of uninformed users. . . . the importance of defining landscape, patch, matrix and landscape context at a scale and in a manner that is relevant and meaningful to the phenomena under consideration.

Nearest neighbour analysis (Thiessen polygons)

Several names exist for what is essentially the same idea, including Voronoi polygons and Dirichlet tessellations. Thiessen polygons divide a region into a series of polygons, such that each polygon contains a single measured point. The polygons are constructed so that they are as compact as possible forming nearest neighbour regions: that is, any location inside a polygon is closer (in terms of distance) to the location of the measured point in the centre than to any other measured point. Thiessen polygons are sometimes used as a form of interpolation; however, for most phenomena there are usually better schemes. Thiessen polygons are useful for allocating variables to points; for example, the points might represent individual plants and the polygon the space available for the roots. Figure 5.2 provides a worked example looking at the space around flowering orchid stems.

Testing for a random distribution of points

It is often important in ecological studies to know whether a set of points is randomly distributed across space. Unfortunately for the dispassionate researcher it is very difficult to 'see' randomness, because your brain will

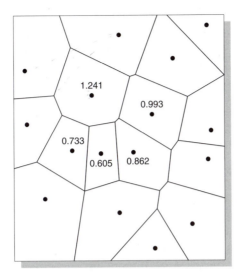

Figure 5.2   Thiessen polygons.

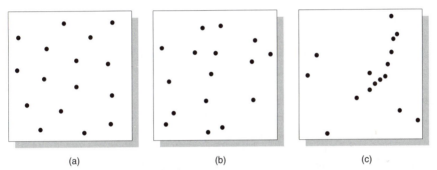

(a)   (b)   (c)

Figure 5.3   Three distributions of points.

always try to see a pattern or detect clusters in a set of data. For example, you might be investigating the distribution of individual orchid plants on a site. Your first idea might be: 'because of competition for scarce resources the plants in a stand will be too regularly spaced to be random'. Your second idea might be: 'if the stand is the result of a single initial individual spreading seeds they will be too clustered to be random'. So, which idea is better or is the distribution spatially random?

Figure 5.3 shows three sets of points; before reading further decide for yourself the distribution: (a) regular, (b) random, or (c) clustered.

The initial test for the distribution of events should be against *complete spatial randomness* (CSR). CSR can be simulated by using pairs of random numbers from a uniform distribution; these pairs represent locations such as x,y (or easting, northing) points. Such a series of points will follow a *homogeneous Poisson process*. The important property of a Poisson process is that if successive samples of sub-regions are taken the mean number of events will be equal to the variance; this property forms the basis of the simple test shown in Box 5.2.

Box 5.2   Simple test for complete spatial randomness

1. Choose some size of 'quadrat' to define sub-areas in the region of interest.
2. Randomly (or regularly) place the quadrat in the region and count the number of 'events' that occur in the quadrat. Repeat this process N times (N should be at least eight). Beware of biasing the results with overlaps between placement of quadrats or by having too many quadrats cross the boundary of the region.
3. Calculate the mean number of events, $\bar{x}$; if the mean is less than 1.0 increase the size of the quadrat and repeat step 2. (Empirical evidence suggests that a mean of 1.6 is preferable.)
4. Calculate $\chi^2 = \Sigma(x_i - \bar{x})^2/\bar{x}$
5. Compare calculated χ^2 with tabulated values of chi squared with $N-1$ degrees of freedom (χ^2_{N-1}), as found in standard statistical texts.

Note that 'quadrat' methods to determine whether a pattern is random depend critically on choosing an appropriate quadrat size: depending on the scale you are using the same pattern might appear to be random, dispersed or clustered. A second 'problem' with quadrat methods is that typically the most important information they convey is that you need to do some more analysis!

For a Poisson process the variance equals the mean; therefore if the variance is greater than the mean this indicates too much variation to be random (clustering). If the variance is less than the mean there is not enough variation and the pattern is too regular to be random. The index of cluster size (ICS) is calculated as $(s^2/x - 1)$; a value greater than zero indicates clustering, a value less than zero indicates regularity.

An alternative and more graphical approach to testing for randomness is to estimate the 'dispersion' of points relative to a number of threshold distances. For each point we can calculate the distance 'near_neighb' to the nearest neighbouring point and a distance 'near_bound' to the nearest boundary. For some arbitrarily selected threshold distance we can count the number of points where 'near_bound' is greater than 'threshold'. For the points where 'near_bound' is greater than 'threshold' we can count the number of points where 'near_neighb' is less than 'threshold'. The ratio of these two counts has a value between zero and 1.0:

$$R(\text{threshold}) = \frac{-(\text{'near\_bound'} \geq \text{'threshold'} \geq \text{'near\_neighb'})}{-(\text{'near\_bound'} \geq \text{'threshold'})}$$

This process is repeated for a number of different threshold distances. Repeatedly plotting R(threshold) against threshold produces a graph giving information about the spatial pattern. If all the points were regularly spaced (like trees in an orchard) increasing the threshold distance from a small value up to the planting distance will have little effect on the 'counts'; once the threshold value equals the planting distance the 'counts' will change rapidly, but thereafter there will be no change. Alternatively, if the points are clustered

Table 5.4 Typical searching questions

Typical questions	Sub-class of operation	Section
I have collected some elevation data across my study site; what is the elevation of my permanent quadrats?	Interpolation	B.1
I have collected some random samples of earthworm biomass; where should I go to collect more samples to provide a better estimate of where high densities occur?		
What data do we have on the occurrence of species within the site?	Search by region	B.2
How much woodland is there within 100 m of the edge of the site?		
Which plant species occur on the ungrazed swards growing on calcareous loam soil type?	Search by attribute	B.3
Where are the sites where the pollution deposition level is greater than 50 μm^{-2}?		

the 'counts' will change rapidly for small threshold distances; as the threshold distance gets large most points will have been 'counted' and changes in the ratio will slow down. A random distribution of points will have a wide range of separation distances, so the ratio R will change steadily as the threshold increases. A graph that rises steeply and then levels off implies local clustering, while if the line starts off gently and then climbs steeply it implies regularity. Wu *et al.* (1987) provide a method to generate random, dispersed and clustered patterns that may be needed to test the 'null hypothesis' of some of the more sophisticated tests described in Bailey and Gatrell (1995).

Section B: using a GIS to search spatial data

Searching procedures fall into the fundamental class of questions (Table 5.4): what is the value at a particular point (interpolation), what is within a particular location (search by region), and where is a particular feature (search by attribute)?

Section B.1: interpolation

Interpolation (in a GIS sense) is the process of estimating values at locations where you have no measurements. It is important to realise that not all phenomena can be interpolated. If the value at a location is completely independent (uncorrelated) of its neighbours then an interpolated value has no meaning. Unfortunately very few people explicitly check that spatial autocorrelation exists before they start to interpolate. Box 5.3 provides a glossary of the jargon.

Box 5.3   Glossary of interpolation terms

autocorrelation – (sometimes *spatial* autocorrelation) values are influenced by the surrounding values; an essential property of phenomena that can be interpolated

anisotropy – changes in the property being investigated is not independent of direction

covariance – the degree with which points close together tend to be similar

covariogram – (also covariance function) the relationship between the **covariance** and the separation of pairs of points

heteroscedasticity – variance of errors is not constant over the whole region

isotropy – changes in the property being investigated is independent of direction

stationarity – arbitrarily selected sub-regions will have similar values for the same property (mean and variance are independent of location)

variogram – (strictly semi-variogram) the relationship between the variance between points and their separation

Any interpolation technique makes more or less strict assumptions about the behaviour of the phenomena being studied and the quality of the information at the locations where measurements have been taken. A simple but unrealistic set of assumptions is that:

1. the known points are well distributed to capture the behaviour of the phenomena,
2. there are no significant errors at the measured points, and
3. the values of the phenomena change in a consistent manner between adjacent points.

Thus:

- Assumption 1: if data are being collected for a topographic surface it is possible to have 'well-distributed' sample locations. The samples can capture the 'extreme' events (hill tops, hollows, river valleys) and places where the phenomena change abruptly (cliff edges, changes in slope) because the surveyor can perceive the phenomena qualitatively and use that information to decide where to collect the quantitative data. Unfortunately there are few ecological phenomena where the spatial changes can be perceived qualitatively so that sampling (measurement) points can be located to capture all the significant variations and changes. Most ecological data arise from systematic, random or stratified random sampling; even though the data collection may be *statistically robust* there is no certainty that such data are adequately located to describe the *spatial* behaviour of the phenomena.
- Assumption 2: with the exception of kriging, responsibility for accounting for uncertainty in the measured values is left with the person doing the interpolation.
- Assumption 3: all interpolation techniques assume that the phenomena change in a consistent manner between adjacent points: they differ in the best way to describe these changes (discrete, continuous, linear, polynomial, statistical and so on).

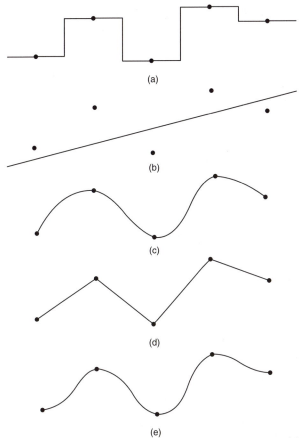

(a)

(b)

(c)

(d)

(e)

Figure 5.4 Typical cross-sections generated by interpolation techniques: (a) Theissen polygons, (b) trend surface, (c) spline, (d) TIN, and (e) IDW (kriging similar).

It is sometimes difficult to visualise what assumptions the different interpolation techniques are using. Figure 5.4 shows 'typical' cross-sections across a surface generated by a number of common interpolation techniques. The techniques can be classified into stepped surfaces, exact surfaces, analytic surfaces and statistical surfaces.

- Stepped surfaces: nearest neighbour allocations (Theissen polygons) all change; is assumed to take place at the edges of homogeneous polygons, and the edges where change takes place are as far as possible from the measured points.
- Exact surfaces: triangular irregular networks (TINs) assume a linear change in values between points. The surface is made of triangular facets.
- Global analytical surfaces: a plane or polynomial surface is fitted through the observations in an analogous manner to fitting a line through a graph, usually using the least squares criterion.
- Local analytical surfaces: a spline fits a curved surface to successive sub-sets of the points. The surface is locally smooth, an exact spline is constrained to go through all the data points, but maxima and minima need not occur at measured points.

- Statistical surfaces: the interpolated surface is the weighted mean of the observed values. Different weighting schemes exist; inverse distance weighting just uses the distances between points, while kriging uses a more sophisticated scheme based on the arrangement of points.

Common GIS interpolation schemes include inverse distance weighting (IDW), triangular irregular networks (TINs), kriging, surface fitting and regression.

Triangular irregular networks

Given a set of locations it is possible to connect all of them so that the whole area is divided up into triangles. Obviously there are many possible ways of joining any set of points so that each point is the **vertex** of one or more triangles. An optimal arrangement is to have each triangle as compact as possible; a Delaunay triangulation is one where for all triangles a circle passing through the three triangle corners will not contain any other point. Once the area is divided into triangles the interpolated value at any location is calculated by assuming that it lies on the plane of the appropriate triangle. TINs can be constructed that incorporate abrupt changes, by specifying 'break lines' to represent 'cliffs' or edges across which triangles cannot be constructed.

The edge of the TIN is bounded by a **convex hull**. This is the shortest line that can encompass all the points, and values are only interpolated within the convex hull. Note there are sometimes reasons why the TIN should not be extended up to the edge of the hull: for example, all that data that you collected on the density of epiphytes on coconut plantations on Tahiti should not be interpolated across bays in the coastline (sorry, just day-dreaming – anyway it is probably too hot and humid to do any field work there at the moment). The convex hull has other uses than bounding a TIN; a common use is to identify the 'core' area of a distribution by successively removing points on the hull and then creating a new hull.

TINs are often used as the basis for contouring routines (where they are easily identifiable because the resultant contours consist of short straight-line segments; see Figure 5.5).

Inverse distance weighting

Given a set of measured values it is assumed that the value at the location of interest is going to be more similar to the points close by than the points far away. The interpolated value is the weighted mean of the neighbouring (measured) values. Inverse distance weighting (IDW) uses some function of 1.0 divided by the distance as the appropriate weight. Typically the weights are the square of the inverse distance ($w_i = 1/d_i^2$). The higher the power to which the distance is raised the less influence distant points have on the result. IDW produces a relatively smooth surface and interpolated values will tend to be closer to the mean value than the measured values. It is possible to limit the number of points included in the interpolation to the nearest, which introduces some 'regional' effect, but it is not possible to change the weights from $1/d^2$ to, say, $1/d^3$ in different parts of the same interpolation.

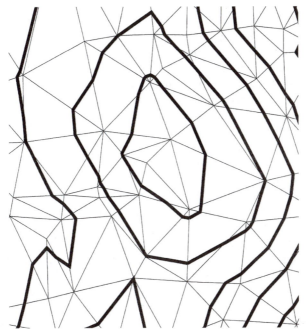

Figure 5.5 Using a TIN for contouring.

Trend surfaces

Surface fitting is analogous to line fitting through a scatter of points on a graph. In some sense trend surfaces remove 'noise' that distracts the eye from seeing regional patterns. However, a trend surface is virtually *useless* for local predictions. It is possible to fit plane, polynomial or trigonometric surfaces as trends. Fitting is usually done with linear regression or simple least squares. Unless there is some very good reason to suppose that your data set has a polynomial or trigonometric trend it would be wise to restrict surface fitting to a linear surface or a very simple curve, and use other techniques to interpolate complex local responses.

There are several technical problems with surface fitting; in particular, outliers can influence the whole surface, clusters of data points mean that the distribution of points is rarely wholly satisfactory, and errors are rarely completely spatially independent. It may sometimes be beneficial to transform data before fitting the surface with strongly positively skewed data; taking logarithms may help. If you knew the structure of the **covariance** of the residuals you could use generalised least squares – but there is no way to estimate it from the observations.

The interpolation technique 'kriging' (described below) assumes that there is no trend in the data, so simple surface fitting is sometimes required as a first stage and for the truly pedantic be part of an iterative process.

Kriging

Kriging could be seen as an elaboration of the IDW scheme. Instead of a single arbitrary weighting function applied to all distances, the weighting function is allowed to vary depending on the data itself; because of this kriging is sometimes referred to as an optimal technique. Kriging relies on being able to fit an appropriate model to the **variogram**. A variogram is simply a graph that relates how similar points that are close together are compared with points which are far apart. Points that are close to each other should have a low variance, and as the distance between points increases so should the variance; at some point (threshold) the variance reaches a maximum and any increase in separation will not increase the variance. Several 'models' exist to describe the increase in variance with distance – always inspect the variogram before making your choice. Once the covarience has been estimated interpolation takes place as for IDW except that the weights are taken from the variogram model.

An important property of the kriging process is that clusters of points do not have an undue influence on the resulting surface. For successful kriging you need at least 50 data points, although of course a GIS will fit a surface with far fewer. A very useful by-product of kriging is information on the spatial distribution of uncertainty in the interpolation, and this can be used to provide guidance on where more samples could profitably be taken as well as how much confidence in the result can be given. There are various extensions and elaborations to simple (universal) kriging, all beyond the scope of this book (an interested reader is referred to Isaaks and Srivastava, 1989, or Bailey and Gatrell, 1995).

Section B.2: search by region

One of the simplest and in a way most powerful capabilities of a GIS is the ability to extract data for a particular region. The region may be defined in many different ways but typically it is a polygon or one or more raster grid cells. The accuracy of the search will be influenced by the precision (and accuracy) with which the boundary is defined and the accuracy of the location of the points of interest.

Search by region using raster data is straightforward; using vector data it is simple when the region is defined as a polygon. If the 'region' is defined by a point or a line it will usually be advisable to construct a buffer around the point or line first. The simplest search by region would be for points wholly within a region; however, it might be necessary to search for features that are partly within the region, or that cross the region or that are near. Figure 5.6 shows the differences in selecting vectors (representing streams) that are wholly or partly within a polygon (representing a particular site). Fortunately visual inspection is usually sufficient to determine which option is more appropriate.

Search by region with raster data is free of many of the complications of searches with vector data. The only issue that may give rise to problems would

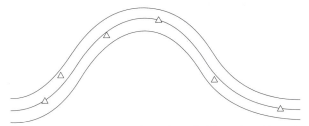

Figure 5.6   Illustration of a search by region using a buffer to 'catch' nearby points (points and lines need 'buffering' before searching).

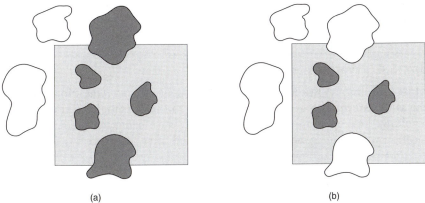

(a)   (b)

Figure 5.7   Illustration of a search by region: (a) through or within the buffer, and (b) within the buffer.

be with data stored at different resolutions, where it may be necessary to resample one data set to a finer resolution before buffering.

A buffer may be developed around any type of feature: point, line or area. With vector data it is usual to produce a coverage that defines whether a point is inside or outside the buffer zone (Figure 5.7). With raster data it is common to have an implicit distance from the edge of the feature. A raster buffer may be viewed as a very simple form of 'cost surface' (see Section D.4).

Section B.3: search by attribute

A search by region could be thought of as a special case of search by attribute: the case where the attribute represents geographic space. Because a GIS contains a database it is very easy to extract information about a particular point, line, cell or object. At its simplest this may merely be the desire to extract and plot, say, all the lakes or roads. At a more complex level it may be a desire to identify areas that satisfy a particular set of criteria.

Section C: using a GIS to classify spatial data ────────

Typical classification questions are given in Table 5.5.

Table 5.5 Typical classification questions

Typical questions	Sub-class of operation	Section
What combinations of soil, climate and plant communities occur in this region? How much territory do species A and B occupy in common?	Overlay	C.1
What are the 'typical' or distinguishable combinations of soil, climate and plant communities?	Classification	C.2
What is the statistical relationship between soil pH, slope, aspect, grazing pressure and the occurrence of species X?	Multivariate statistics	C.3

Box 5.4   Warning – no data

Most GIS are very strict about the representation of *no data*. No data is *not* the same as *zero*. If a quantity is zero then you really should store it as such. Typically any operation that results in no data being accessed at a given location will result in no data at that location after the process is completed.

Section C.1: overlay

Overlays, together with interpolation and classifications, are the archetypal GIS operators. Manual overlays were widely promoted in the ground-breaking text *Design with Nature* (McHarg, 1969). Suitability maps were constructed by stacking many single-theme maps, each drawn on transparent material on a light table. The resultant map would be very dark where several constraints coincided and bright where there were few. As a result of the success of manual overlays the technique was very rapidly computerised, which improved the number of layers that could be dealt with. Computers also allow the 'weight' or importance of each constraint to be varied quickly and easily.

Overlays can be carried out using raster or vector data. Overlays typically come in two basic 'flavours', which can be termed 'logical' and 'mathematical', although the distinction can be somewhat blurred. Logical overlays imply the use of if . . . then . . . else constructs with AND, OR and XOR operators. Mathematical overlays use arithmetic, trigonometric and similar functions. You need to check how your GIS represents various operations; for example, greater than might be represented by '>' or 'gt', not equal by '! =' or 'ne' or '<>'. Note also that normally a data layer cannot be on both sides of the equals sign (a=a*a would be acceptable in many computer languages but in a GIS you would normally have to do: b=a*a; kill a; a=b; kill b). Be aware that data will sometimes (always?) contain exceptions that you have not thought of; a GIS will often replace the results of dividing by zero, and logarithms of negative numbers by entering a *no data* value (but try not to rely on it).

Examples and implementations of overlay procedures are given in Boxes 5.5 and 5.6, respectively.

Box 5.5   Examples of overlay procedures

Logical overlays

if (*statement that can be evaluated as either* true *or* false) then statement
  {further optional statements}
{else {optional test}
  statement}
{else further optional tests and statements}

```
(i)    if (A gt B) then output = 1
(ii)   if ((A gt B) AND (C ne 0 or D eq 3))
       then
         output = 1
       else if (E lt D) then
         output = 2
       else
         output = 3
(iii)  if (((rainfall lt 700) AND (temperature gt
       20)) AND
       ((soil_class eq 3) OR (land_class eq 1)))
       then suitability = 1
```

Mathematical overlays

new data layer = function (some data layer(s))

```
(iv) A = B + C
(v)  A = B + EXP(C * 0.1 + 0.5) + TAN(D)
(vi) number_of_flowers = number_of_flowers_last_year +
       march_temperature * 0.1 - SQRT(june_rainfall)
```

Box 5.6   Implementations of overlay procedure

GIS	Command
ArcInfo	`new_map = map1 * map2` `if (map1 gt 0 and map2 gt 0) new_map = 1`
GRASS	`r.mapcalc new_map = 'map1 * map2'` `r.mapcalc new_map = 'if (map1 > 0 && map2 > 0)'`
Idrisi	`overlay x 3 map1 map2 new_map` {no exact correspondence to 'logical' overlay; need to use reclass and then use arithmetical overlay}

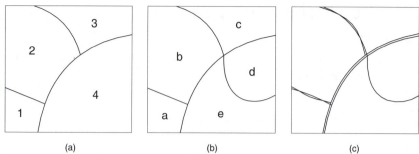

(a)   (b)   (c)

Figure 5.8 Problems of polygon overlay: (a) layer 1, with four polygons; (b) layer 2, with five polygons; and (c) overlay, with 12 polygons.

When overlaying vector data a common problem is the production of numerous '**sliver polygons**', which result from minor differences in the representation of the boundaries. Paradoxically the more closely the boundaries on the two layers follow each other the more numerous the sliver polygons. Figure 5.8 shows two vector map layers representing, say, land cover and soil type. Layer 1 has four polygons and layer 2 five polygons; however, after the overlay procedure there are 12 polygons! Of course the big question with this overlay is how many of the 12 polygons represent significant features (combinations) and how many are artifacts (sliver polygons) of the data collection and representation process. The simplest, most honest approach is to do nothing about the exponential increase in the number of polygons. However, to improve the aesthetic appearance or simplify the analysis it is quite common to reduce the number of polygons arbitrarily based on size or shape. What you do (if anything) will partly depend on whether you are motivated to try and understand the relationship between vegetation and soil in the study area or whether you need a pretty map for your report.

Section C.2: classification

An ecologist needs to take care with the use of the term 'classification' when using a GIS. It can be used in the sense of *allocating* one class into another as well as in the more 'conventional' sense of organising multivariate data.

Classification functions included in most GIS are most usually designed for the classification of remotely sensed data. Maximum likelihood and iso-mean classification techniques are the most common techniques used. Classification methods for remotely sensed data are usually available for supervised and unsupervised methods. With supervised classification homogeneous patches of each class are identified on the image, and these patches are used to provide the information needed to classify the rest of the image. In unsupervised classification the algorithm identifies the classes on the basis of the spectral response and the user them tries to identify what each class is.

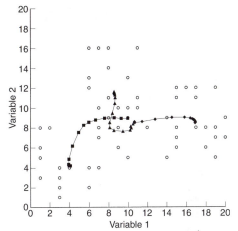

Figure 5.9   Classification process: fuzzy clustering algorithm: (◆) centre 1, (■) centre 2, (▲) centre 3, and (○) data.

The general idea of a 'fuzzy' classification can be illustrated graphically. In the simplest case, two variables such as reflectance values in two distinct bands in the spectrum are used as co-ordinates to show the distribution of values. Figure 5.9 shows the process with a limited data set (60 points). The user selects the number of classes to be generated (in this case three); with this algorithm points are allocated at random to a group so the centroids start out very close to each other. An iterative process is started, where the distance from each point to the centre of each group is calculated; these distances form the basis of an estimate of 'membership' of each point to each group. Each point is then allocated to the 'nearest' group, and a new centroid for that group calculated. After 25 iterations (with this data set) the position of the centre of each group cannot be significantly improved. The distance calculated is usually the Euclidian distance, but other metrics can be used to account for differences in the statistical distributions of the variables. One of the attractive aspects of this classification scheme is that the membership is calculated for all groups, so some estimate of how reliably a point has been calculated can be made. You need to remember of course that clustering techniques always produce the required number of groups; it is your responsibility to make sure they have any meaning.

Reclassification

Reclassification is one of the archetypal GIS operations; it is also one of the quickest, easiest, most efficient and most powerful ways to destroy information! Reclassification is typically used to merge classes or convert continuous values into discrete classes; for example, you might want to merge coniferous and broad-leaved woodland into a single class – 'trees' – or to classify areas where rainfall is below some value as one class and all other areas into a

	A	B .	C	D	E	F	G
1	54.7	58.3	0.6823	m	h	h	m
2	61.7	51.7		m	h	h	m
3	65.5	39.5		l	m	m	l
4	52.8	57.3		l	l	l	l
5	53.3	57.6					
6	60.9	60.4		m	m	l	m
7	62.5	66.7		m	h	h	m
8	52.4	57.1		l	m	m	l
9	39.1	38.3		m	l	l	m
10	55.1	58.5					
11	49.2	55.6					
12	39.1	38.3					
13	37.9	41.9					
14	39.3	38.9					
15	39.3	38.9					
16	37.2	41.2					

Figure 5.10   Fragment of a spreadsheet illustrating regression experiment.

different class. Reclassification can be accomplished using if . . . then . . . else constructs (for more details see Section C.1), but the logic can rapidly become difficult. A more manageable technique is to use *lookup tables* or *remap tables*. Both can be used to allocate single classes to new classes or ranges of values in discrete classes.

Section C.3: statistical analysis

Univariate statistics: a GIS experiment without a GIS

Regression is a widely used tool in statistics, but the experiment below shows how misleading it can be when applied in a simple-minded way to spatial data. As an added extra the same experiment also demonstrates some of the problems with classifying data. The example makes use of a spreadsheet so you can try the experiment before you get your GIS. Figure 5.10 shows a fragment of a spreadsheet that holds data for two maps, A and B. Before they were rounded both sets of values had the same mean (50), standard deviation (10) and skewness (0). Areas D1–G4 and D6–G9 display the two 'maps' after they have been classified. The contents of the 'map cells' are a series of 'if' statements that specify what 'class' to use; in this example 'h' is greater than the mean plus one standard deviation and 'l' is less than the mean minus one standard deviation; that is:

```
if (a1 > 60,'h', if (a1 < 40, 'l', 'm'))
```

The contents of cell C1 is the Pearson correlation coefficient between the values in columns A and B. For 16 pairs of values (14 degrees of freedom) and $\alpha = 0.01$ the critical value for r is 0.6226 and at $\alpha = 0.001$ the critical value

Series A	Series B	Series C					
			Map A				
50.6	60.7	50.6		m	h	m	m
61.8	42.0	58.4		m	h	h	m
59.5	49.9	70.3		l	h	h	l
43.1	59.1	48.8		m	m	l	m
43.3	59.5	49.3					
61.0	64.3	57.4	Map B				
61.0	53.7	59.4		h	m	m	m
42.9	58.7	48.4		m	h	m	m
37.6	40.1	36.6		m	h	m	m
61.0	61.0	51.0		l	m	m	l
61.0	55.4	45.5					
36.1	40.1	31.2	Map C				
57.1	37.9	61.0		m	m	h	m
41.8	40.5	45.5		m	m	m	m
37.6	40.5	36.8		l	m	m	l
44.6	36.6	49.7		h	m	l	m

Figure 5.11   Three statistically identical distributions.

of r is 0.7420, so for the initial set of data we can be between 99 and 99.9% certain that there is a positive linear correlation between the two sets of numbers.

So, compare the two 'maps'; would you say they looked very similar? Now you can start your experiment: change a couple of numbers in column B and see what effect this has on the correlation coefficient and on the maps. The linear correlation coefficient is usually considered to be fairly robust. Figure 5.11 shows three other possible data sets (all with a mean of 50, a standard deviation of 10 and a skewness close to zero); note that in each case the statistical significance of all three is *exactly* the same, $r = 0.6823$.

While you can produce similar patterns by trial and error the examples in Figure 5.11 were produced using the linear programming facility in the spreadsheet: by relaxing the skewness and deviation constraints you can produce very different maps with very similar correlations.

Multivariate statistics

Many of the multivariate statistics that are included in GIS packages are there to help classify remotely sensed data. For more complex statistical analysis it may be more appropriate (quicker, more reliable) to dump the data into an external statistical package (such as SPSS, Minitab, Splus, etc.) and then re-import the data. Divisive classification algorithms are rarely found in GIS packages.

Some GIS allow for the construction of 'stacks' of data layers, which allow the analysis of multiple pairs of data sets. Figure 5.12 shows a set of multiple scatter graphs from a stack of data.

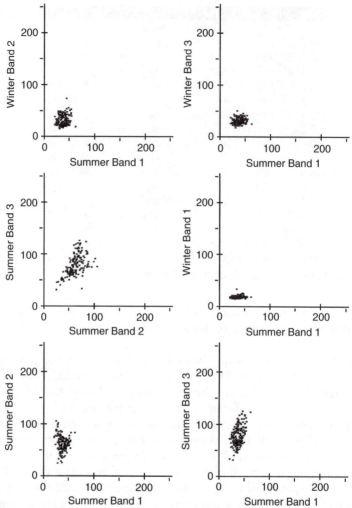

Figure 5.12   Examples of 'small multiples' to display multivariate data.

Section D: using a GIS to model spatial data

Typical modelling questions are given in Table 5.6.

Whatever form a model takes and however it is implemented or developed, you always need to bear in mind the difference between the relative simplicity of what is mathematically tractable and the rich complexity and uncertainty in real ecological systems.

Section D.1: network models

A network model in a GIS consists of a series of points (or nodes) connected by lines (or links). Ease of movement along the links can be specified as an

Table 5.6 Typical modelling questions

Typical question	Sub-class of operation	Section
If a seed falls into a river at point X and it remains viable for N days, where might it manage to establish? Is it possible for species A to get from X to Y without intervention?	Networks	D.1
How many sources of emigration are in the neighbourhood of each site?	Neighbourhood modelling	D.2
What is the probability that species X occurs at location Y?	Statistical modelling and Bayesian inference	D.3
What is the easiest (least expensive) way for species A to get across a heterogeneous landscape?	Cost–diffusion–spread	D.4
How will a wild fire spread across this forested landscape?	Map algebra	D.5

impedance, which can vary depending on the direction of movement. Additional impedances can be applied at junctions so that movement between different links can be more or less easily achieved. Examples of networks in GIS are often concerned with human transport networks; however, the process works well with ecological processes that are strictly confined to a network. We have interfaced a hydrodynamic model with a GIS network model to describe the movement of floating and submerged seeds along rivers (Wadsworth *et al.*, 1997).

Section D.2: neighbourhood modelling

Neighbourhood modelling means altering the value at a location depending on the values in the neighbouring cells (there is no real reason why you should not apply the same ideas to polygon data, but it is not so common). The neighbourhood may be defined by a moving 'window' or by 'zones'. In either case the neighbourhood may be just the immediately adjacent cells or it may include those that are close but not adjacent. A moving window does not have to be symmetrical (but they usually are). A zone is typically an irregular area that has been defined by an additional data set.

Section D.3: statistical modelling and Bayesian inference

Standard statistical methods such as regression and correlation can be applied (with care) to spatial data, and some GIS also include facilities to help make inferences through Bayesian statistics. Beware! Attitudes vary towards Bayesian statistics; before you start to use it check the attitude of your supervisor.

Bayesian estimation of average values

Empirical Bayesian estimation is concerned with whether a value estimated from local (regional) data provides a better estimation of the true value than a value estimated from all available data. Suppose that the number of grouse infected by internal parasites has been estimated for a number of different moors. Within any region the observed rate is equal to the number infected divided by the number of birds sampled, or $r_i = y_i/n_i$. Suppose that we had only managed to sample a few birds in a particular moor. We might then suppose that our observed rate of infection was not very reliable; alternatively if we had caught a lot of birds we might suppose our estimate was more reliable. There are several approaches to producing a Bayesian estimation that can be adopted; the simplest (and the only one we will discuss) is that the mean rate of infection, μ, and its standard deviation, σ, are *aspatial* and should be estimated from the entire data set for all regions using the method of moments. We can now 'weight' our observed rate of infection to obtain a 'Bayesian' estimate of the rate, B_i:

$$B_i = w_i \times r_i + (1 - w_i) \times \mu$$

where:

$$w_i = \sigma/(\sigma + \mu/n_i)$$

It follows that if n_i is very large (that is we caught lots of birds), w_i will tend towards 1 and B_i will tend towards r_i, because we think our observed value is robust. If n_i is small then w_i will tend to be small and our best estimate will tend towards the 'global' mean value, μ; if we fail to catch any birds in a region then w_i would tend to zero and B_i would equal μ.

Bayesian estimation of species distributions

An alternative formulation of a Bayesian model can be used to predict the distributions of species. Suppose that you wish to estimate the presence (or absence) of a species in a mosaic of landscape habitats. In Baysian terms:

$$P(P|H) = \frac{P(P) \times P(H|P)}{P(P) \times P(H|P) + P(A) \times P(H|A)}$$

Where $P(P|H)$ is the final or *posterior probability* of the occurrence of the species if the habitat is present; $P(P)$ is the initial or *prior probability* of the occurrence of the species; $P(A) = 1 - P(P)$. $P(H|P)$ is the *conditional probability* that the habitat is present if the species is present, and $P(H|A)$ is the *conditional probability* that the habitat is present if the species is absent.

An interested reader is referred to Tucker *et al.* (1997) for a detailed description of the practical issues in using such equations in a GIS environment. Problems arise from the spatial resolution of available data, mobility of species being studied, need to use 'surrogate' variables for values not measured, availability of data to validate the model, relation between habitat variables and other environmental variables, and sensitivity of the model.

Regression and correlation

Regression and correlation can be applied (with care) to spatial data. The reason you need to be careful is that the technique is aspatial and does not take into account the spatial *pattern*. One useful technique is to perform a regression analysis using the same data twice but displacing them a given distance when used as the dependent variable. A high correlation from displaced data shows that the landscape has a distinct pattern at the scale of the displacement. Autocorrelation also indicates phenomena that are suitable for interpolation. Logistic regression is often included as an option for regression.

Section D.4: cost surface, distribution and neighbourhood models

Cost–diffusion–spread

'Cost–diffusion' and 'spread models' are concerned with converting geographic distance into a cost–distance (usually on a raster grid). Distance is the simplest measure of the cost of getting between selected points. As the environment is heterogeneous the cost of crossing some areas will be higher than for other areas. Provided some estimate of 'quality' or cost can be assigned to each cell then the cumulative cost from any point is the sum of the distance across all the cells multiplied by the cost of each one. Such surfaces can then be used to find optimal (least-cost paths) across the landscape. Such approaches have potential applications in landscape ecology, to study the energetic costs of dispersal or migration in relation to measures of landscape 'resistance'.

One problem with using these techniques in ecology is likely to be the need to put quantitative estimates on qualitative judgements in an attempt to view the landscape in a particular way. It is very easy to guess what is important for the dispersal of a species, and one supposes very easy to get it wrong. Plate 1 provides an example of finding a suitable corridor or optimal path.

Gravity modelling

Gravity modelling is concerned with the flow of 'individuals' between sources and sinks (or origins and destinations). It is widely used in commerce to try and identify who is likely to go and shop at which stores (especially because everyone wants to run down the centre of our inner cities and make sure only people who like to drive have access to decent shops). Gravity models are primarily *descriptive* in nature. The number of individuals travelling from the ith source to the jth destination is the product of the propensity of that origin to generate flows, the attractiveness of the destination and some function of the distance between the two. For a 'Newtonian' gravity model the distance function is proportional to the inverse square of the separation. There are possible applications in population ecology to study relationships between availability of feeding and breeding resources and the distribution of individuals. Much of the literature on gravity modelling is in the social sciences, but see for example Haynes and Fotheringham (1984).

Section D.5: map algebra

Most GIS now contain features to allow fairly complex models to be constructed using the basic procedural operators to allow choice ('if' statements) and repetition ('do', 'for' and 'while' loops). Models may operate to call particular functions and to operate on individual features or raster cells. Models written in a GIS shell or script are typically interpreted rather than compiled and therefore likely to be several orders of magnitude slower than a model written in C, Fortran or Pascal; on the other hand you have direct access to lots of display functions and statistical functions.

As an example of how to use map algebra a simulation of competition between an annual and a perennial plant species has been devised. Below are three realisations of the simulation (Boxes 5.7–5.9) using ArcInfo's macro language (AML), a UNIX shell script using GRASS commands and the Idrisi macro language (IML). As with any simulation there are many different formulations of the problem and there are compromises between efficiency and clarity to be made. The rules of this simulation are very simple: the perennial species is dominant, and during each time step a perennial can spread into any adjacent unoccupied cells. During each time step perennial plants die at random with a user-specified probability. Cells occupied by the annual species spread seeds a certain number of cells in all directions; if a cell is unoccupied by a perennial plant and it has seeds in it (from the previous time step) an annual plant will establish and set seeds. Simulations use a mortality rate of 10% and a seed rain that reaches three cells in all directions and should result in the persistent existence of the annual species. With a spread rate of two cells the annual species usually goes extinct within 50 generations; extinction is usually accompanied by what look like groups or clusters that are for a while apparently persistent.

Example of simulation using ArcInfo's AML

Some explanation of the AML is required, it uses the GIS functions:

- `expand`   to grow out a specified number of cells from a specified 'class',
- `rand()`   which fills a grid with random numbers between zero and one,
- `if, else`   construct which operates on each corresponding cell in several grids.

In addition to the 'data analysis' functions the AML uses the general functions to specify the area of interest and display the data:

- `setwindow`   specifies where data should be calculated,
- `setcell`   specifies the size of a grid cell,
- `mapex(tent)`   what part of the grid should be seen,
- `image`   which is the simplest of the commands to show a grid,
- `kill`   to delete a grid.

Finally the AML commands are prefaced with &. This macro uses:

- `&sv`   to set the value of a constant,
- `[exists]`   to see if <name> of <type> already exists,

- `%<variable>%` to return the value of the variable,
- `&if &then` to control the flow of calculations; multiple statements in `&do &end` block,
- `/*` to specify a comment.

Box 5.7 Example of map algebra using ESRI's ArcMacro language

tested: Arc version 7.0.2, Sun Ultra 1

```
/* competition.aml
/*
/* An AML to simulate competition between a
  perennial and an
/* annual species (see text for specific details)
/*
/* decide on an 'area' of interest
setwindow 0 0 100 100
setcell 1
mapex 0 0 100 100
/*
/* 'User'-defined variables:
/* mortality = probability of a perennial cell
  dying
/* spread = how far an annual can spread its
  seeds
/* generations = number of generations to simulate
&sv mortality := 0.1
&sv spread := 3
&sv generations := 50
/*
/* kill off any grids that might be lying around
  from previous runs
&if [exists perennial -grid] &then kill perennial
  all
&if [exists annual -grid] &then kill annual all
&if [exists seed -grid] &then kill seed all
/*
/* generate an initial pattern - half perennial
  and half annual
tmp = rand()
if (tmp ge 0.5)
  begin
    perennial = 1
    annual = 0
  end
```

(continued)

(continued)

```
else
  begin
    perennial = 0
    annual = 1
  end
endif
kill tmp all
/*
/* see what the initial pattern looks like
image annual
/*
/* generate a 'seed rain' around the annual plants
seed = expand (annual, %spread%, LIST, 1)
/*
/* simulate a fixed number of generations
&do I := 1 &to %generations%
  /* let the perennial expand one cell
  p_growth = expand (perennial, 1, LIST, 1)
  /* generate a random probability
  p_surv = rand()
  kill perennial all
  if (p_growth gt 0 and p_surv gt %mortality%)
    perennial = 1
  else perennial = 0
  /* tidy up
  kill p_growth all
  kill p_surv all
  /*
  /* let the annual plants exploit any empty cells
    which have seeds kill annual all
  if (seed gt 0 and perennial lt 1) annual = 1
  else annual = 0
  /*
  /* generate a new 'seed rain'
  kill seed all
  seed = expand (annual, %spread%, list, 1)
  /*
  /* see what the new pattern looks like
  image annual
&end
&return
/*
/* A few ways to make the simulation more
  realistic/complex
```

(continued)

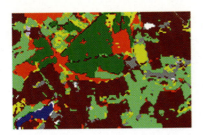

Land Cover

Reclassed as costs
(blue = cheap, green = expensive)

A

Cumulative Costs from A
(black = low value, white = high value)

B

A

Cumulative Costs from B

B

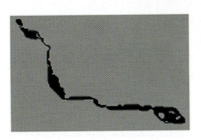

Corridor
Black = within 3% of cheapest possible route

Plate 1. Optimum corridors

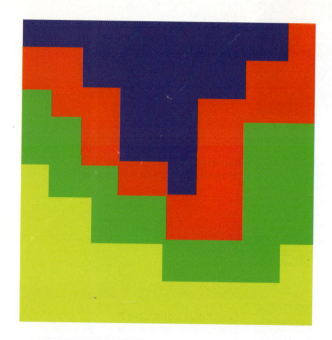

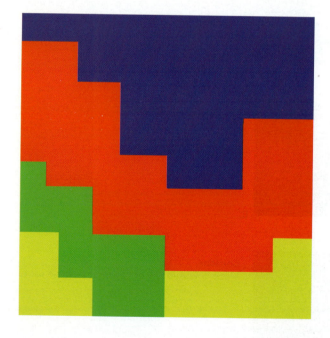

Plate 2. Equal area and Equal interval maps
See table 15 for details

Total Ecosystem area (km²)

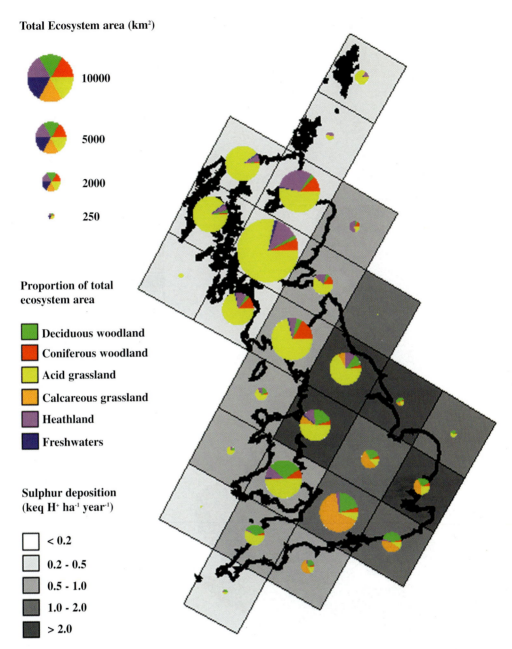

10000

5000

2000

250

Proportion of total ecosystem area

- Deciduous woodland
- Coniferous woodland
- Acid grassland
- Calcareous grassland
- Heathland
- Freshwaters

Sulphur deposition (keq H⁺ ha⁻¹ year⁻¹)

< 0.2

0.2 - 0.5

0.5 - 1.0

1.0 - 2.0

> 2.0

Critical Loads Mapping and Data Centre, ITE Monks Wood, February 1997

Plate 3. Proportional ecosystem areas and EMEP 1990 sulphur deposition
(Data acknowledgement: CLAG soils sub-group, CLAG freshwaters sub-group,
CLAG vegetation sub-group, ITE Bush, AEA Technology, IH)

Inland Water
Beach and Coastal Bare
Saltmarsh
Grass Heath
Mown/Grazed Turf
Meadow/Verge/Semi-natural
Rough/Marsh Grass
Moorland Grass
Open Shrub Moor
Dense Shrub Moor
Bracken
Dense Shrub Heath
Scrub/Orchard
Deciduous Woodland
Coniferous Woodland
Upland Bog
Tilled Land
Ruderal Weed
Suburban/Rural Development
Continuous Urban
Inland Bare ground
Felled Forest
Lowland Bog
Open Shrub Heath

Plate 4. Land cover map of Great Britain

```
(continued)
/*
/* a) devise a suitability map for each species
/* b) make the perennial mortality rate stochastic
/* c) make the 'seed rain' stochastic, (spatial
   and/or temporal variation)
/* d) introduce regular or random catastrophic
   events
/* e) allow seeds to survive longer than one year
/* f) change occupancy from a Boolean (true/false)
   variable to a continuous
/* variable
/* g) only let the perennial plants spread after a
   given number of generations
/* h) introduce more species
```

An alternative implementation of the same problem using a UNIX shell and the GRASS GIS is shown in Box 5.8 (the sh shell is preferred but csh usually works fine). Some explanation of the shell is required: commands starting `r.` are raster functions, `g.` are general functions and `d.` are display functions. The main GIS functions are:

- `r.buffer`   to grow out a specified number of cells (`distances=X`) from a specified 'class' (in this case the default of all non-zero cells),
- `r.random`   to create random 'sites'; in this case fill X% of a grid with 1s, leave the rest as 0,
- `r.mapcalc` to carry out a piece of map algebra,
- the if statement has the format, if (expression that can be evaluated as true: result if true, result if false). If the expression is numeric zero is defined as false; anything else as true.

GRASS protects data by location so that the temporary grids created in the simulation can be overwritten 'on the fly' and do not have to be deleted.

In addition to the 'data analysis' functions, the shell uses two other GIS functions:

- `g.region` gives north, south, east, west limits and north–south, east–west cell resolution,
- `d.rast`   which is the simplest of the commands to show a grid.

Finally the shell uses the sh syntax:

- `<variable> = <value>`   to assign a value (note that spaces around the = sign will confuse things,
- `$<variable>`   to return the value of the variable,
- while – do loop to control the flow of calculations; the counter is updated with `j = 'expr $j + 1'`,
- `#`   to specify a comment.

Box 5.8 Example of map algebra using GRASS

tested: Sun Ultra and GRASS 4.1

```sh
#!/bin/sh
#
# competition.sh
#
# shell to attempt to simulate competition between
  an annual and
# perennial species (see text for details)
#
# decide on an area of interest
g.region n=100 s=0 w=0 e=100 nsres=1 ewres=1
r.mapcalc tmp=1

# Variables which define the dynamics of a
  simulation
# mortality = probability of a perennial dying
# spread = how far an annual can spread its seeds
# generations = number of generations to simulate
mortality=10
spread=3
generations=50

# generate an initial pattern - 50% perennial 50%
  annual
r.random input=tmp nsites=50% raster_output
  =perennial
r.mapcalc annual='if(perennial,0,1)'

# see what the initial pattern looks like
d.rast annual

# spread the seeds around from the annual plants
r.buffer input=annual output=seed distances
  =$spread

# simulate a given number of generations
j=1
while [$j!=$generations]
do
  # let the perennial expand its territory and let
    some die
  r.buffer input=perennial output=p_growth
    distances=1
  r. random input=perennial nsites=$mortality%
```

(continued)

```
(continued)
raster_output=p_surv
  r.mapcalc perennial = 'if (p_growth > 0 &&
    p_surv < 1, 1, 0)'

  # annual plants exploit any empty cells that
    have seeds
  r.mapcalc annual = 'if (seed > 0 && perennial <
    1, 1, 0)'

  # spread a new set of seeds
  r.buffer input=annual output=seed
    distances=$spread

  # see what the new pattern of annual looks like
  d.rast annual

  j='expr $j + 1'
done
# see Box 5.7 for ways to make this simulation
  more complex
```

The final implementation of the problem uses the Idrisi macro language (IML) and is shown in Box 5.9. Some explanation is required: IML has no facility to do loops or perform logical switches. Therefore repeated actions require repeated lines. The main GIS functions are:

- initial   defines a uniform grid (which is used in the costgrow function) and defines the area of interest,
- costgrow   is used here to create a buffer,
- random   creates a random grid (values between zero and 1),
- reclass   changes the values in a grid according to the rules in the text files *.rcl,
- overlay   used here to multiply two grids together.

In addition to the 'data analysis' functions IML uses:

- display   displays a grid,
- delete   deletes a grid,
- branch   calls the second IML.

Finally:

- rem   specifies a comment.

Box 5.9   Example of map algebra using the Idrisi macro language

(needs to be written as four text files)
tested: Idrisi for Windows version 1

file 1 – competition.iml

```
rem competition.iml

rem Idrisi macro to simulate competition between
  perennial and
rem annual species (see text for details)

rem generate a uniform field to define area (used
  later on)
initial x unif 2 1 1 2 100 100 plane m 1 100 1
  100

rem generate initial random pattern (of zeros and
  ones)
random x peren 1 1 unif 1 1

rem generate the inverse pattern to represent the
  annual plants
reclass x I peren annual 3 inverse

rem have a look at the initial pattern
display x a annual cols2 N x 0

rem repeat the simulation for as many years as you
  require
branch grow
branch grow
branch grow
branch grow
branch grow
```

file 2 – grow.iml

```
rem grow.iml

rem allow the perennial plants to spread by one
  cell
costgrow x peren unif 1 per2 2
delete x peren.img
delete x peren.doc
```

(continued)

(continued)

```
rem generate a random field with values between 1
  & 100
random x rand 1 1 unif 1 100
reclass x I rand surv 3 mortal
rem combine expanded population with random
  mortality
overlay x 3 surv per2 peren
delete x rand.img
delete x rand.doc
delete x surv.img
delete x surv.doc

rem allow seeds to spread up to 3 cells from the
  initial locations
costgrow x annual unif 3 seed 2
delete x annual.img
delete x annual.doc

rem allow annuals to grow in any unoccupied cells
  which have seeds
reclass x I peren open 3 inverse
overlay x 3 open seed annual
delete x open.img
delete x open.doc
delete x seed.img
delete x seed.doc

rem see what the new distribution looks like
display x a annual cols2 N x 0
```

file 3 – inverse.rcl

```
1 0 0
0 1 1
-9999
rem changes zeros to ones and vice versa.
```

file 4 – mortality.rcl

```
1 0 90
0 90 100
-9999
rem changes values between zero and 90 to have a
  value of 1
rem changes values between 90 and 100 to zero
```

Note that if you have access to more than one of the systems you will not get exactly the same response from different simulations; hopefully this is just due to different random number generators. When you type in either of these simulations neither will run the first time; we suspect that this is because of a typing error (of course if we sell enough of these books some clever person will get it right first time). Quote marks and spaces are the most common problems; for example, `&sv spread := 3`, `&sv spread=3` and `&s spread 3` will all work but `&sv spread=3` probably will not. Similarly, the line `annual = 0` will work but `annual :=0` will not work.

Review

We have tried to introduce you to some of the facilities in a GIS that we have found to be of use in ecological studies and made an attempt to provide some structure to the large number of commands by classifying them as being concerned with measuring, searching, classifying and modelling data. However, there are many other functions and facilities of interest to the ecologist which will emerge as you become more experienced.

Further reading

For a rigorous but readable introduction to spatial analysis we recommend T.C. Bailey and A.C. Gatrell *Interactive Spatial Data Analysis* (Longman, Harlow, 1995), they also include a computer disk of Info-Map which allows several types of spatial analysis not generally included in a GIS, with the book. Specific advice on using ArcInfo to carry out the point pattern analysis can be found in A.C. Gatrell and B.S. Rowlingson 'Spatial statistical modelling within a GIS framework', in A.S. Fotheringham and P. Rogerson (eds) *Spatial Analysis and GIS* (Taylor & Francis, London, 1993). Several GIS vendors produce their own texts on GIS in addition to the user manuals; for example: *Understanding GIS: The ArcInfo Method* (Longman, Harlow) or J.R. Eastman *IDRISI: A Geographic Information System* (Clarke Laboratories for Cartographic Technology and Geographic Analysis, Clarke University, Worcester, Mass., 01610–1477, 1993). A variety of GIS tutorials are available on line: the USGS produces an introduction to GIS on **http://info.er.usgs.gov/research/gis/title/html**, while a 'typical' software-specific introduction can be found at **http://hermes.inrs-urb.uquebec.ca/~boivinj.acinfo.html** (note this particular example is in French for ArcInfo – the Internet is global).

Chapter 6

Visualisation and communication

Visualisation and communication

Having collected your data, imported it into the GIS, cleaned it up, checked for errors, carried out your analysis – what more can there be? While the quest for truth for its own sake may be laudable, there are very few of us who do not need to communicate the results of any analysis. Bear in mind that even if you are only communicating the results to *yourself* (seeking understanding) you still need to be able to communicate effectively and efficiently. Communication of spatial data will almost always involve the production of some form of map. As a guide to portraying quantitative information we strongly recommend Tufte (1983); even though many of the examples he uses are aspatial, the process of thinking about communicating information is exemplary.

Tufte's five principles of graphical excellence

1. Graphical excellence is the well-designed presentation of interesting data – a matter of substance, of statistics and of design.

2. Graphical excellence consists of complex ideas communicated with clarity, precision and efficiency.

3. Graphical excellence is that which gives the viewer the greatest number of ideas in the shortest time with the least ink in the smallest space.

4. Graphical excellence is nearly always multivariate.

5. Graphical excellence requires telling the truth about the data.

Often the distinction between data exploration and data visualisation becomes blurred. The basic visualisation tool in spatial analysis is the map. Production of maps has been greatly speeded up by computers – while a cartographic-quality product always requires long and careful design, simple (and sometimes truly ugly) maps can be produced in a few minutes. It is not necessarily the aim always to produce cartographically excellent maps, but in the last few hundred years cartographers have battled with many of the problems of how to present spatial information within the limits of the physiology and psychology of perception. It would be arrogant (and foolish) to ignore the experience of so many cartographers working for so many years.

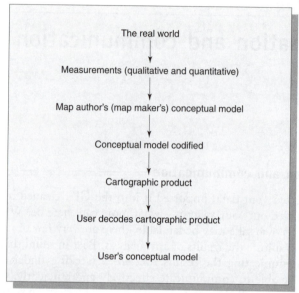

The real world

↓

Measurements (qualitative and quantitative)

↓

Map author's (map maker's) conceptual model

↓

Conceptual model codified

↓

Cartographic product

↓

User decodes cartographic product

↓

User's conceptual model

Figure 6.1 Visualisation and communication (from a coding–decoding perspective).

There is a considerable literature describing the cartographic process in terms of coding and decoding and information theory; unfortunately it is an activity that generates more and more uncertainty the more you think about it. Consider the flow chart in Figure 6.1. It starts with the lovely complex real world; we then take some measurements, which help in the construction of a *conceptual model* of some part or aspect of the world; we then formalise the conceptual model by a graphical code consisting of a set of lines and colours on a flat surface; finally, the map reader decodes the colours and squiggles to form his/her own conceptual model, which, hopefully, bears some resemblance to the real world. We all know the process works, and often works well, but why does such a complex process work at all?

One of the problems with viewing the use of maps as a coding–decoding process is that it implies that each element on a map is a discrete element. The presence of a single contour line tells us that a particular value of one variable (elevation) exists at those points, but a collection of contours tells us about the shape of the surface, from which we can infer many other things. On many maps it is not even clear what a single element of the code means: a thin black line might mean a field boundary or the edge of a road. It is only the *relationship* with other equally unidentifiable black lines that allows us to know which it is and to build up our conceptual model of the landscape.

Mapping and the 'truth'

The process of coding information into a cartographic product and then reading the code produces a paradox: to be a useful product a map represents the real world by telling 'white lies'. Features are combined, exaggerated, displaced, ignored or represented by symbols in an attempt to convey a 'true' picture of the world. For many decades producing maps was solely the province of large national mapping organisations (and a few large commercial organisations), but these organisations tended to produce consistently reliable products so that the naïve user has tended to take maps on trust. Now that we can all produce maps a more cynical view needs to be taken. Monmonier (1991) is fast becoming the standard work on the use and abuse of maps and is strongly recommended as an introduction to how to spot the more common forms of deception (we assume, of course, that you do not want to perpetrate such crimes!). The critical point that Monmonier makes is that any map is just one representation out of an *infinite* number of possible representations.

The main problem with the graphical representation of information is that most education systems are so strongly biased towards verbal communication. Society rewards verbal manipulation and dexterity. Professionals who exhibit it, such as lawyers, advertising executives, salesmen and politicians, are afforded respect (and more toys than ecologists) although sometimes the respect given is grudging and accompanied by feelings of ill-will. On the other hand most of us stop using pictures to tell a story (or convey information) at a very young age and outside of Japan the 'graphic novel' (or comic book) is not highly regarded.

Graphical elements of maps

Fortunately most ecologists will be working at such a large scale that many of the more complex issues of cartographic design can be neglected; it is, however, useful to consider some of the aspects of basic cartography. The main graphical elements of any map are symbols, lines and colour.

Symbols

When selecting a symbol to use there are several properties that need to be considered. These are the ability of the map reader to *detect* the symbol, to *discriminate* one symbol from another, and to *identify* the symbol. These properties can be broken down further so that identification conveys the need for *recognition* and for *interpretation*, while discrimination has both qualitative and quantitative aspects.

Lines

Lines may represent features that are identifiable on the ground (such as a field boundary) or concepts (such as a contour or isoline). The use of copper-plate engraving for printing meant that for many years maps consisted entirely of lines, with perhaps some colour wash applied by hand for the 'de luxe' editions. Even today many maps such as road atlases consist primarily of lines.

Box 6.1 Problems with the use of colour to represent a continuous variable (adapted from Keates, 1973)

1. The series of colours must form a *progressive* series in which the steps are evenly balanced but all perceptibly different. These differences must be perceptible for small areas yet not overpowering for large areas.
2. Whatever design is used, it will influence all the other elements of the map to some degree.
3. Associated information may be concentrated within certain 'colour zones'. For example, most human activity (cities, roads, agriculture) takes place at low elevations.
4. The change from one colour to another is an abrupt visual change that emphasises the division on which the classification is based. Yet the change in the value of the surface (such as elevation) may be imperceptible; the greater the contrast between adjacent colours, the greater the apparent interruption of the continuous surface into a series of apparently separate stepped surfaces.
5. The choice of intervals decides the units that will be shown. The classification of the surface into zones cannot be modified locally in relation to the most important changes of elevation within a given area. These arbitrary changes can produce a very misleading impression of the characteristics of particular regions.

Colour

Colour is widely used as a quick way to make an impact. Unfortunately the impression you make by liberal or unconsidered use of colour may not be the one you intended. The interested reader is referred to any standard text on cartography, but some of the more salient aspects to colour are discussed below.

A number of conventional sequences exist to represent values on a continuous scale. One of the most widely used is the sequence of colours used to represent elevations (**hypsometric** tints); such a sequence can easily be confused with colours used to represent vegetation. Keates (1973) identifies a set of five problems with using colour to describe variations of a continuous surface, specifically elevation, but similar arguments apply to any continuously varying surface; these problems are shown in Box 6.1.

Anyone who has spent a happy (or frustrating) afternoon trying to get a sequence of colours just right will appreciate why a few conventional sequences are used so often. The final problem with the use of colour is the significant proportion of the population who have difficulty with distinguishing some colours. Probably everyone at some time tries to use the full spectral sequence: violet, blue, blue-green, green, yellow, orange, red. It usually turns out to be extraordinarily difficult to associate such a scale with quantities or even decide what the order is; for example, is green higher or lower than yellow? A single sequence of a particular colour from a light shade through to an intense shade of that colour, such as very light red, pale red, medium red,

red, deep red, is the colourful variation of a grey scale. Single-sequence schemes are usually easy to associate with a range of values, but note that you will probably still be limited to only six or seven easily distinguishable shades. The most complex colour scheme that (sometimes) works is the part-spectral single-sequence arrangement, such as pale yellow, light yellow, light orange, orange-red, red. If the range of values includes both negative and positive then a double-ended multi-hue scale sometimes works. You might decide to use reds for negative values (as they do in accounting) and blues for positive values – but beware, your reader might associate blue with cold (and negative values) and red with heat (and positive values).

Less than 1% of women and about 8% of men suffer from some form of impaired colour vision; the two most common types are *protanopes*, who have impaired vision at the red end of the spectrum, and *deuteranopes*, who have problems with the green part of the spectrum. If you *have* to use colour and *have* to make the map readable by *everyone* then Brewer (1996) provides a review of a number of experiments into non-confusing colour combinations. Brewer identifies three issues causing confusion with colours: naming, visual impairment and simultaneous contrast. Colour pairs that are accurately read by all map users are red–blue, orange–blue and orange–purple (yes let's hear it for those orange and purple maps!); brown–blue is possible and so are yellow–purple and yellow–blue (but you have to be careful with yellow because it is naturally only a light tone).

On computers, individual colours are usually specified as mixtures of three primary colours (red, green, blue), where each primary colour is specified in the range zero to 255. It is very difficult given three numbers to guess what the colour is going to look like – try it as a party game – well it is no odder as a game than charades. You may of course discover that such a scheme produces different colours on different computer screens and that when printed the colours need not bear much relation to what you were expecting. Note that not only does the colour differ between the screen and the paper but also the type of paper can influence whether the reflection is *specular* or *diffuse* (matt or glossy) and whether the paper is opaque or translucent will also affect the colour. Human perception of colour is also influenced by the size of the area printed and what colours are adjacent to each other. The use of colour to distinguish different types of line is complicated by the apparent changes in colour as it crosses different areas. One of the clearest examples of this is by drawing a grey line over a pale colour and then a dark area. Over the pale area the grey will look very dark (black) and over the dark area very light (white). Similarly blue surrounding a grey patch will give the grey a yellowish tint.

Other schemes to specify colours can rely on wavelength: at one limit of human detection is violet with a wavelength of 400 nm; through the spectrum there is blue-green at 500 nm, orange at 600 nm and at the other limit to human detection red at 700 nm. Other methods of specifying colour rely on the hue, saturation (or chroma) and lightness (or value).

Grey scales

Grey scales are not as visually dramatic as colour but at least there is very little room for confusion as to which way the 'scale' is running; either from light to dark or dark to light. Although your software package may tell you it can produce hundreds of shades of grey, it is close to impossible to distinguish more than six or seven shades. Grey shades are often difficult to reproduce; the image may be subtle and elegant but as soon as you photocopy it, it becomes unreadable.

Non–spatial elements of maps

Map scale bar

There are three ways to represent map scales: verbally, with ratio scales and graphically. Verbal scales are such things as 'one inch to the mile' or 'one centimetre represents five hundred metres', while ratio scales are such things as: 1 : 62 500 or 1 : 50 000 or 1 : 1250. Graphical scales exist in many different forms. They have one great advantage over verbal and ratio scales, which is that they are immune to different methods of reproduction: photocopying, on a computer screen, being dragged and manipulated before printing in a report and so on. Verbal and ratio scales should not be used on computer displays and should be used on other media with discretion.

North points

North points are often an opportunity to add embellishment and decoration to a map, and like bar scales they exist in a wide variety of forms. Before you get carried away remember that there are *three* north directions: true north, magnetic north and chart north. True north is aligned with the axis around which the Earth spins, magnetic north wanders around (slowly) in the Arctic. Chart north will be aligned with true north at the reference longitude but, depending on the projection, may or may not be aligned with it elsewhere on the map. We would suggest that you keep the north point symbol as simple as possible and be as explicit as possible about which north it refers to!

Legend

This provides a written description of the symbols used. As Monmonier (1991) remarks 'a legend might make a bad map useful, but it can't make it efficient'.

Map projection

Maps are two-dimensional representations of a complex three-dimensional object (the globe). A map projection is a set of mathematical equations for converting co-ordinates on the spheroid into co-ordinates on a plane surface.

Box 6.2   Glossary of cartographic terms

aliasing – (also called 'jaggies') occurrence of jagged lines on a computer screen or other raster display, caused by approximating a line to a series of cells where the level of detail exceeds the available resolution; most disconcerting when it happens to text

axis – a reference line in a co-ordinate system

Cartesian – a method of representing values by two (or three) axes perpendicular to each other (first devised by one of our heroes, René Descartes (1596–1650))

cartogram – a diagram (rarely used in ecology) where space is deliberately distorted to allow comparison of values between regions with very different spatial extent; the skill is in allowing the user to recognise the regions after the distortion has been applied

choropleth – a map where the phenomena are represented by homogeneous polygons

clip – extraction or display of features within a polygon

dpi – dots per inch – a measure of resolution; the higher the dpi the finer the resolution

GKS – graphical kernel system – a series of specifications that allow graphics to be drawn using Fortran programs

graticule – lines printed on a map to aid a user to locate particular co-ordinates

GUI – graphical user interface – system that allows a user to interact through a mouse (or other pointing device) as well as a keyboard

hue – one of the aspects of colour, dependent on the dominant wavelength?

hypsometric – refers to a sequence of colours used to represent elevation

isopleth – a map consisting of lines joining equal values

legend – a written description of the symbols used

neatline – the border around a map, title, key, etc.

saturation – how intense a colour is (also called chroma)

scale – the relationship between the actual distance between two features and the separation between the representation of the two features on the map

---

Map co-ordinates

Maps typically use Cartesian co-ordinates in a similar manner to those used on graphs. A regular grid or **graticule** is usually printed on the map to aid the location of features (to be pedantic a graticule and grid are not synonymous). The x-axis of the grid is generally termed the eastings, the y-axis the northings. The rule is 'in the door and up the steps', i.e. along the eastings axis (x-axis) and the northings (y-axis). It is typical to estimate locations to one-tenth of the grid spacing. Within the UK each 100 k block has a unique two-letter identification, and some people still make use of these in preference to specifying the 100 km value in numbers. Note also that the grid is oriented in relation to the projection and does not normally line up with magnetic north (the location of which varies from year to year).

Map scale

On his travels Gulliver met a people who produced a map of their territory at a scale of one to one (1 : 1); everyone else represents real-world features smaller on the map than they are in reality. National mapping agencies produce products at a variety of scales; in the UK the Ordnance Survey produces maps at 1 : 1250, 1 : 2500, 1 : 10 000, 1 : 25 000, 1 : 50 000 and so on, but the two largest scales (1 : 1250 and 1 : 2500) are produced only for selected areas. A *large-scale map* needs a *large* piece of paper to represent a site, a *small-scale map* needs a *smaller* piece of paper to represent the same site. A distinction is sometimes made between a *plan* and a *map* on the basis of scale, but a better distinction is on the basis of cartographic representation: a map uses symbols to represent some items; a plan does not.

Map generalisation

An important point to remember about maps is that they are graphical representations of the real world. There are several forms of generalisation: symbolisation, exaggeration, classification, displacement and selection. If we take an example of a 1 : 50 000 scale map then buildings are shown in their correct location but using a stylised *map symbol* for house, church, mill, etc. Roads will be shown in their correct location but with an *exaggerated* width so that their class can be represented (note that an A road and a B road are classified on the basis of the importance of the route not necessarily on physical differences). When it comes to natural features location may not be exact; it is a quite legitimate cartographic practice to *displace* features spatially so that they can be seen clearly. Some environments are characterised by features that individually are too small to be portrayed but *en masse* are important; for example, with an area with lots of small bog pools, the cartographer has to *select* and exaggerate some of the pools so that they can represent the 'feel' of the landscape, although some of the features shown on the map do not exist on the ground. The human element in cartographic design should never be neglected; in open moorland a simple ditch might be displayed because there is nothing else to show, while in a city a river might be ignored because it can make the map look too cluttered. Remember that just because the map is in electronic form it is still a *cartographic* product.

Types of map

There is no shortage of ways to represent data on a map; the problem is deciding which is the best method to use! Common methods are homogeneous coloured patches (choropleth), lines joining equal value (**isopleth**), perspective views and block diagrams, grid maps, vectors, points and symbols, **cartograms**, or some mixture of all of them.

Table 6.1 Example data set to illustrate difference between equal area and equal interval map

Value	Frequency	Class	
		Map 1 – equal area	Map 2 – equal interval
0.2	16	1	1
0.7	8	2	1
1.2	8	2	2
1.7	16	3	2
2.2	4	4	3
2.7	4	4	3
3.2	4	4	4
3.7	4	4	4

The data are distributed below.

0.2	0.2	0.2	0.2	0.2	0.2	0.2	0.7
1.2	1.2	0.2	0.2	0.2	0.2	0.7	0.7
1.7	1.2	1.2	0.2	0.2	0.7	0.7	0.7
1.7	1.7	1.2	0.2	0.2	0.7	1.7	1.7
2.2	1.7	1.7	1.2	0.2	0.7	1.7	1.7
2.7	2.2	1.7	1.7	1.2	1.2	1.7	1.7
3.7	2.7	2.3	2.3	1.7	1.7	1.7	3.2
3.7	3.7	2.7	2.7	3.2	3.2	3.2	3.7

Choropleth maps

These represent the variables by a set of homogeneous polygons. When they are used to describe a continuous surface the way the range is split up can have a great effect on the visual perception of the result. There are two common default methods to split up a range of values: *equal intervals* and *equal area* or *quantile* classes, with an equal number of cases in each class. Refer to the values givenn in Table 6.1. Note that, unrealistically, we have no outlying values in this data set.

With a quantile scheme the data can be mapped as in Plate 2a. Using an equal-interval scheme the map looks like Plate 2b; the question is whether you would realise that it was the same data set. There is of course no universally 'correct' scheme for dividing up a range of values. The honest approach is to prepare a number line or histogram and be explicit about how the range has been split up, and how any 'odd' or extreme values have been dealt with.

Isopleth

The most commonly encountered form of isopleth map is a contour map of elevation. Lines join points with equal values. Please make sure that when you annotate the lines the figures read uphill (some well-known GIS packages have the numbers reading up and downhill at random!).

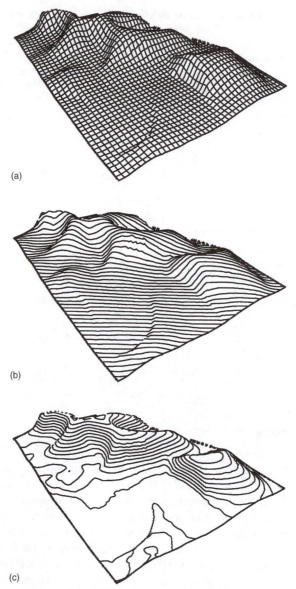

Figure 6.2 Perspective view and block diagrams: (a) mesh or net, (b) profiles, and (c) contours.

Perspective views and block diagrams

These are ways of representing surfaces and solids pictorially. Automatically producing perspective views (Figure 6.2) turns out to be much more difficult than might be expected. As your nice shaded image of data you are familiar with appears on the screen your mind helps you build up the view, but once the image is printed or shown for the first time it almost always causes confusion. To be successful keep your block diagrams as simple as possible and include lots of visual clues such as a draped mesh or lines to inform the viewer that it is a surface. Unfortunately we offer much advice on producing good perspective views of landscapes seen from the ground – because we have never produced one that was truly satisfying.

Gridded maps

Grids are often displayed using colours to represent values (a choropleth map with regular 'polygons'). It is also sometimes possible to portray gridded data as vectors of direction (for example, Figure 5.1) or if the number of cells is small enough with the numeric content of the cells.

Lines, vectors, points and symbols

These are the traditional methods of conveying data on a map. A variety of line and symbol types will be available in your GIS. Note that size is one of the few unambiguous methods for conveying information. One non-traditional method for conveying information is the ability to use 'statistical' graphics with spatial data, for example the use of pie-charts within polygons to show the amount of different types – see Plate 3.

Cartograms

A cartogram is a map where space is distorted. The object of the distortion is to allow for comparison of values between regions with very different spatial extent. The main application of cartograms is in social geography. For example, parliamentary constituencies have approximately the same population, but they vary in area from a few square kilometres (in the centre of a major city) to thousands of square kilometres (in rural areas). If some attribute (say unemployment) is mapped by constituency the image will be dominated by the large sparsely populated constituencies and the small densely populated urban constituencies will be difficult to detect. By distorting space the cartogram allows the different constituencies to be seen. What makes a good cartogram difficult to produce is the need to control the distortion so that the viewer can still recognise that a map has been produced. An ecological application might be to produce a cartogram showing the number of species on each continent – of course, because most people have a very poor understanding of the relative size of the different continents the impact of your map might not be so dramatic.

Designing your own map

Before you start designing your own map there are three questions you need to have considered:

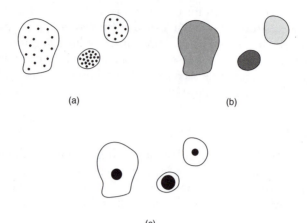

(a)　　　　　　　　　　　　　　　　　　(b)

(c)

Figure 6.3　Three methods of portraying information about point counts: (a) points, (b) shading, and (c) sized symbols (try experimenting before you start).

1. Why is the map being constructed?
2. Who is the map for?
3. How will the map be presented?

You might be constructing the map for yourself so that you can understand the data (in which case you might be able to dispense with some of the niceties). You might be preparing the map for your professor (in which case it had better be perfect). It might be for a journal (in which case you need something monochrome the size of a postage stamp) or it might be for a talk or a poster or any one of a dozen other reasons. It is always worthwhile to carry out a few experiments to test which is the best way to convey the information. Figure 6.3 shows three methods (choropleth, points and symbols) for showing information about flocks of birds feeding on the inter-tidal zone.

The sophistication and skill of the intended audience may have an influence on the design and style of the map, but people who do not respect the intelligence of their audience *never* produce good maps.

Having clarified what the purpose of the map is and who the audience is you next need to decide how it is going to be produced. The main alternatives are computer screen, overhead projector acetate, slide and paper. There are advantages and disadvantages of each of the different media. When maps are primarily designed for display on a computer screen then the order in which the different elements appear can aid the viewer, in particular when the map represents a perspective view. Seeing the surface being created helps to understand the structure being represented. On a more sophisticated level, computers can be used to generate animated sequences of maps representing time series, and they can also be used to select the region of interest interactively. Unfortunately the resolution on computer screens is very poor; a typical good-quality computer screen might have a resolution of 1200 by 800, but now a cheap and cheerful printer has a resolution of 300 dots per inch, so our fancy 19-inch computer screen can display a little less detail than can be printed onto a sheet of paper the size of a postcard.

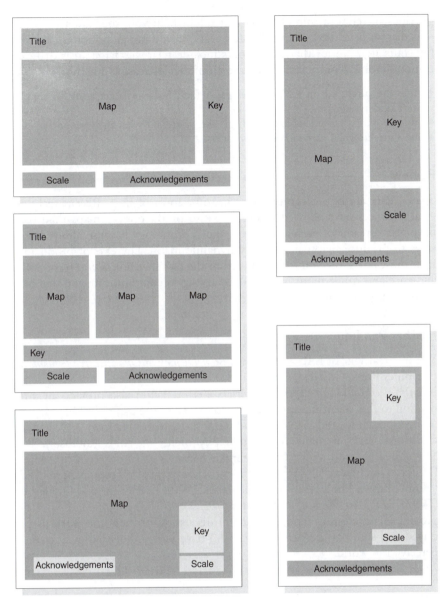

Figure 6.4 Sketches of potential layouts.

The first stage in designing any map is to get a pile of scrap paper and start to sketch out the layout. Sketch out where the primary map graphic is going to be, then where any subsidiary graphics (such as a location map, cross-section, histogram) are going and then where the title, key, scale bar, north arrow, acknowledgements need to be. Layout is a difficult thing to give any precise guidance on, so it might be a good idea to search around for maps that you like and then 'borrow' what you need from them. If you are producing maps for a thesis, report or journal check the style guide first (they sometimes specify titles must be at the bottom, or all text on a separate sheet). Figure 6.4

provides sketches of some layouts that might help you; they are actually layouts that we have used over the last few years (so you can judge our graphic skills for yourself!). Notice that our layouts include relatively large areas dedicated to non-graphical purposes. There are a number of reasons for this:

- most are produced on A4 sheets of paper (and nothing larger than A3 since we were undergraduates),
- most require a considerable amount of text to explain what they are portraying,
- most are designed as stand-alone products.

Review

Spatial data are best communicated in some form of map. Recent technological advances have allowed more people to produce more maps than could have been dreamt of even a few years ago. Technology has also allowed novel ways to produce maps, particularly dynamic maps. The ease of producing a map must not allow you to be blinded to the fact that unless you respect your audience and think about what you are trying to communicate you will never produce a good map.

Further reading

We are great fans of E.R. Tufte *The Visual Display of Quantitative Information* (Graphics Press, Cheslive, Conn., 1983) but R.A. Earnshaw and N. Wiseman *Scientific Visualisation* (Springer-Verlag, Berlin, 1992) also provide an introduction to the 'scientific' approach, as do several of the papers in D.J. Maguire, M.F. Goodchild and D.W. Rhind *GIS* (Longman, Harlow, 1991). More traditional text on maps and cartography including A.H. Robinson, R.D. Sale, J.L. Morrison and P.C. Muehrcke *Elements of Cartography* 5th edition (Wiley, Chichester, 1984) or B. Dent *Cartography: Thematic Map Design* (Wm C Brown Publishers, Dubuque, Iowa, 1990). D. Dorling and D. Fairburn (1997) *Mapping Ways of Representing the World* (Longman, Harlow, 1997) provide an interesting (and up-to-date) discussion on how the world is portrayed and some of the personalities who have been influential in how we all perceive the world.

Chapter 7

Applications in ecology

Introduction

Earlier chapters have demonstrated some of the spatial, analytical and predictive procedures that GIS are designed to facilitate. GIS have a potential role in both theoretical and applied aspects of ecology. In fact it can be argued that GIS have done a lot to bridge the gap between the two. This chapter presents some examples of the effective use of GIS for ecological applications.

By enabling access to new sources of data (especially remotely sensed data) use of GIS has literally opened up whole new areas of ecological research and analysis. It is now possible to study and manage natural resources on a regional level, taking account of processes that were very difficult to study in small geographic areas. Perhaps the most widespread application of GIS in ecology is for mapping and classification of ecological 'resources'. By developing consistent approaches to ecosystem classification and incorporating ecosystem maps into GIS, integration with other resource inventories is facilitated and baselines of ecological information can be established that greatly improve ecological analysis and decision making. Digital map layers are also straightforward to query and can be used not only to ask *where* something is, but to speculate about *why* it might be there and not elsewhere. GIS can therefore be very helpful as a tool in conceptualising problems or hypotheses, particularly in landscape ecology. Field ecologists have been relatively slow to adopt GIS, but this has been partly because of the difficulties of collecting spatially explicit field data. There are now some very good examples of GIS that have combined spatially referenced field data with data derived from other sources, including satellite imagery. GIS can also play a part in identifying the key relationships that might merit further field investigation or in helping to ensure that data collection programmes 'capture' variation at a suitable scale and are based on an appropriate intensity and range of sampling for the desired application.

While GIS have a potential role in many aspects of applied ecology, their commercial use has been relatively restricted. A review of the use of GIS for environmental impact assessment (EIA) by environmental consultants (Joao and Fonseca, 1996) confirmed that GIS are used more for presentation of data and results than they are for manipulation of data or analysis. Even the more simple analytical tools provided by GIS have, to a large extent, been

neglected by practitioners, for example map overlaying to derive simple measurements of habitat area and distribution. Often, GIS have been used simply
to generate 'base maps' that can be used to structure field surveys and provide
a platform for the presentation of data and results. Nevertheless, GIS have
enabled practitioners to look 'beyond the factory fence' and to draw meaningful conclusions about the wider implications of development proposals. This
is partly because GIS have opened access to sources of digital data that were
previously inaccessible (remotely sensed data in particular) and partly because
they permit the use of information from a variety of sources on a common
interpretive platform.

New sources of data

The ability to acquire remotely sensed data has made it considerably easier to
carry out ecological studies with wide geographical coverage. Per unit area,
remotely sensed data are generally much cheaper to acquire than the results of
field surveys. Satellite imagery used for ecological applications has, predomin-
antly, been obtained from the Landsat Multispectral Scanner (MSS), Thematic
Mapper (TM) or SPOT High Resolution Visible (HRV) instruments and has
been used for surveying and characterising terrestrial land use or vegetation
cover. Combined with other data sets in a GIS, remotely sensed data can be
interpreted to distinguish different vegetation cover types more clearly, or to
explain or predict distributions of associated species (Veitch *et al.*, 1995).

In the UK, the Institute of Terrestrial Ecology (ITE) produced a digital
Land Cover Map of Great Britain (Fuller *et al.*, 1994) using classified multi-
temporal Landsat TM imagery. This was the first such map of the UK to be
produced in digital form and was also the first national land cover map that had
been published since the 1930s, when the results of a national land use survey
were published (Stamp, 1962). The Landsat Thematic Mapper was chosen as
a source of data because it includes a detector sensitive to the middle-infrared
(MIR) wavelengths needed to separate a wide range of vegetative and non-
vegetative land cover types. Other potential sources, (notably the SPOT HRV
sensor) might have had a greater spatial resolution, but would not have been
so useful for distinguishing land cover types. The map was derived from TM
wavebands 3 (red), 4 (near-infrared) and 5 (infrared) due to their characteristic
responses from vegetation and also their ability to penetrate haze, a perennial
problem in the UK (Fuller and Parsell, 1990). Landsat data were supplied as
digital scenes of 185 km². It was possible to classify 12% of Britain's land
cover using either winter or summer images alone, but combined winter and
summer images were needed to distinguish between seasonally bare arable land
and land with permanent vegetation like grassland or to distinguish between
evergreen and deciduous woodlands. Locational errors can occur when using
combined imagery. In this case, the summer data were geometrically registered
to the British National Grid using control points on 1 : 50 000 Ordnance Survey
maps and the winter data were re-sampled to fit the 'corrected' summer data.

At the end of the day, locational errors in the *Land Cover Map of Great Britain* were considered to be less than one pixel on average.

To produce the *Land Cover Map of Great Britain*, the satellite imagery had to be classified into distinct land cover types or classes. This was done using the results of a consultative exercise involving end-users to derive suitable categories and then carrying out field reconnaissance surveys to identify land cover in selected land–water 'parcels' or 'training areas' per Landsat scene. Results from these 'training areas' were effectively used to calibrate the classification. Versions of the map were produced using 25 or 17 land cover classes. The 25 classes are described in Table 7.1. Plate 4 shows the '25-class' version, aggregated to show the dominant (most common) type in each 1 km cell. It is valuable for studies that require information at a 'field scale' and, being digital, is readily integrated into GIS. It has been used for a wide range of applications in wildlife conservation, landscape planning, environmental impact assessment and climate change modelling.

For ecological studies, where information on habitat is required, as opposed to information on land cover (often a direct function of land use), the *Land Cover Map of Great Britain* can be overlaid with other data sets to derive habitat distributions. For example, heathland land cover has been extracted from the map and overlaid with data on soils, geology, altitude and the distributions of lowland heathland species to derive potential distribution maps for lowland heathland habitat (Treweek *et al.*, in press). The *Land Cover Map of Great Britain* is based on raster data, and is therefore not directly suitable for certain applications packages that work with vector data structures, without modification.

A 1 km summary data set can also be derived from the map, which has useful applications in national land use planning. The UK Department of the Environment commissioned the Countryside Information System (CIS) to provide policy makers and advisors with access to up-to-date information about the countryside. The 1 km summary of the *Land Cover Map of Great Britain* is one of the key data sets included in the CIS. Other data sets include the results of a national countryside survey carried out by the Institute of Terrestrial Ecology in 1990 and due to be repeated in the year 2000. Useful comparisons can be made at the 1 km scale between land cover information and data on agriculture, wildlife species distributions or any sample digital data set that can be expressed on a 1 km^2 framework. For example, data currently available in CIS format include:

- administrative boundaries,
- hydrological data,
- topographical data,
- soils,
- designated areas,
- farm type,
- wildlife records, and
- critical loads for pollutants.

Table 7.1 ITE *Land cover map of Great Britain* – class descriptions

Class No.	Class title	Description
1	Sea and estuary	Coastal waters and estuaries
2	Inland water	Inland fresh and estuarine waters
3	Beach and coastal bare ground	Inter-tidal mud, sand, rocks, shingle, sand dunes, bare rocks
4	Salt marsh	Inter-tidal seaweed beds, salt marshes up to normal levels of high water spring tides
5	Grass heath	Semi-natural grasslands of dunes, heaths and lowland–upland margins
6	Mown or grazed turf	Improved pasture and amenity swards forming a turf throughout the growing season
7	Meadow–verge–semi-natural grass	Low-intensity management and cropped swards – not maintained as a short turf
8	Rough marsh grass	Lowland grasslands, mostly uncropped – mainly perennial spp., with high winter litter content – not cropped or grazed
9	Moorland grass	Montane–hill grasslands, mostly unenclosed moorland
10	Open shrub moor	Upland dwarf shrub–grass moorland – may be burned in cycles
11	Dense shrub moor	Upland evergreen dwarf shrub dominated moorland – may be burned in cycles
12	Bracken	Bracken-dominated plant communities
13	Dense shrub heath	Lowland evergreen shrub-dominated heathland
14	Scrub–orchard	Deciduous scrub and orchards
15	Deciduous woodland	Deciduous broadleaved and mixed woodlands
16	Coniferous woodland	Coniferous and broadleaved evergreen trees
17	Upland bog	Upland herbaceous wetlands
18	Tilled land	Arable and other seasonally or temporarily bare ground
19	Ruderal weed	Ruderal weeds colonising natural and man-made bare ground – may include brushwood and rough grass
20	Suburban–rural development	Developed land comprising buildings and roads but with some cover of permanent vegetation
21	Continuous urban	Industrial, urban and other development, lacking permanent vegetation
22	Inland bare ground	Ground without vegetation, or surfaced with imported gravel or sand
23	Felled forest	Felled forest with ruderal weeds and rough grass
24	Lowland bog	Lowland herbaceous wetlands
25	Open shrub heath	Lowland dwarf shrub–grass heathland
0	Unclassified	Land that does not fit into the 25 categories above (usually cloud or shadow, etc.)

A full environmental catalogue of data sets available in CIS format is available, and it can be downloaded from **http://www.nmw.ac.uk/ite/cisflier.html**. You can also develop and incorporate your own 1 km grid data sets. The CIS is designed to generate maps and statistics describing the state of a given environment, to identify areas that share a given environmental characteristic, or combination of characteristics, to compare areas exposed to different types of impact or to link ecological data with models of the rural environment to help assess the ecological consequences of land use or other environmental changes. The CIS is based around map and data 'windows'. 'Regions' can be defined for querying data sets using county, regional or country boundaries, grid references, land classes or other 'user-defined' criteria for selection. Once a 'region' has been defined, a data set can be selected and expressed in the form of 'census' or 'sample' data types. 'Census' represents data by generalising them to 1 km grid squares, while 'sample' might involve the averaging of a data set to predict the likely land use in unsurveyed areas, for example (Pipes, 1997).

An example of a task made simple using a system like the CIS is comparing the number and types of sites of special scientific interest (SSSI) found in different counties or regions. The Highlands Region of Scotland actually has the largest number of SSSIs. By combining the SSSI data set with land cover data, it is possible to determine that 'managed grassland' is the dominant land category for SSSIs overall, with obvious implications for prescribing or agreeing appropriate long-term management (Pipes, 1997). Theoretically, using land cover and species distribution data together with other qualifying data sets, it would also be possible to identify which habitat types are over- or under-represented in the total set of SSSIs and therefore to determine whether or not current approaches to SSSI designation are effective and even-handed in terms of protecting the full range of UK wildlife habitats.

The main strength of the CIS lies in its capacity to increase awareness of and access to key digital environmental data sets (Pipes, 1997). This is regarded as a worthy and useful aim, not only at the national level, but more globally also. The Statistical Office of the European Communities, for example, is currently developing a GIS (GISCO) to help in developing regional policy with respect to agriculture, the environment, transport and other sectors. As well as administrative boundaries, details of infrastructure (airports, ports, roads, railways), data on hydrology, altitude, climate, soils, land cover, biotopes and areas designated for nature conservation are being built into GISCO's reference database. Data held on land and nature resources relevant to ecological studies and their resolution include the data sets summarised in Table 7.2.

A study of the impact of trans-European (transport) networks (TENs) on nature conservation was carried out by the Royal Society for the Protection of Birds (RSPB)/BirdLife International using such pan-European data sets, though in this case data were actually derived from data sets held by BirdLife International itself and the World Conservation Monitoring Centre (Bina, 1995). A pilot GIS-based method for strategic environmental assessment (SEA) was explored using data on proposed road and rail networks together with data on

Table 7.2 GISCO's data on land resources and nature resources (as in April 1997)

Data theme or layer	Description	Resolution or scale	Spatial extent
Climate	19 climatic variables for 5308 stations	Location of station	EU 12 (except DDR)
Land cover	Inventory of biophysical land cover in 44 classes	1 : 100 000	EU 12 (44% complete)
Soils	Soil mapping units and their characteristics (15 843 polygons)	1 : 1 000 000	EU 12
Biotopes	Inventory of sites of major importance for nature conservation – 7740 sites at present with information on site identification, site location, site description, ecological information	Location of centre point of site	EU 12 + Finland
Biogeographical zones	Delineation of five different biogeographic zones as defined in the framework of the Council Directive 92/43/EEC	1 : 20 000 000	EU 12
Designated areas	Inventory of sites designated under community legislation and international conventions – 1812 points with information on site identification, location and description	Location of centre point of designated area	Pan-Europe and North Africa
Landscape types	30 European landscape types in eight landscape complexes	1 : 6 000 000	Pan-Europe
Natural potential vegetation	232 vegetation types in 4160 polygons	1 : 3 000 000	Pan-European (except Eastern European countries at present)

the distributions of important designated sites for nature conservation, namely important bird areas (IBAs) and nationally designated sites (NDSs). Corridors 2 or 10 km wide were constructed as buffers either side of planned trans-European routes and the number and surface area of important designated sites falling within them was calculated. The centre points of 21% of the IBAs and 11% of the total surface area of NDSs were found to lie withing 10 km of planned roads or railways, while 4% of all IBAs and 2% of all NDSs were within the 2 km buffer or corridor. This preliminary study highlighted the potential for proposed trans-European transport networks to cause significant disturbance to important European nature conservation sites. Clearly further study would be necessary to explore potential impacts in more depth, but at least it was possible to estimate the potential magnitude of the problem before planning more detailed studies.

Together, remote sensing and GIS have therefore opened up a whole range of new data sources for ecological study. However, the development of national data sets and maps is not, by itself, enough to solve pressing land use and ecological problems. It is invariably necessary to go further and actually to interpret available data; for example, to move on from descriptions of 'land cover' or 'land use' to descriptions of habitat distribution and quality or ecosystem status based on multiple data sets. This is where GIS come into their own.

Ecosystem classification and mapping

Mapping ecosystems requires classification of available, relevant data to derive homogeneous map units with predictable characteristics. It is unlikely that you will ever be able to characterise an ecosystem or map its distribution using just one data set. For example, the 'land base' might be stratified into map units based on information about climate, geology, soils and vegetation. These map units can be displayed as polygons, the data (or attributes) associated with each polygon being stored in a map database. Digital maps generated from such databases provide a spatial expression of ecosystem classifications that can be used to depict habitat, wildlife and other ecological resources in a standardised and directly comparable fashion. Ecosystem mapping is invariably greatly enhanced by drawing on a number of data sources to stratify the landscape. Satellite imagery, aerial photography or radar images may be suitable to derive distribution maps, as well as the results of field survey. Even colour infrared photography can be useful where separation of vegetation structure is required, for example to distinguish species' feeding or reproductive guilds.

There are a number of national and regional programmes for the mapping and classification of 'ecological resources' that are structured and managed on a 'GIS platform'. Perhaps the best known is the USA's ecological monitoring and assessment program (EMAP). The USA's Environmental Protection Agency (US EPA) initiated EMAP in 1988 to provide information on the status, extent and condition of ecological resources throughout the USA. EMAP covers seven broad 'resource categories': near-coastal waters, the Great Lakes, surface waters, wetlands, forests, arid lands and agro-ecosystems, a co-ordinated monitoring network and series of indicator measurements being developed independently for each category (Novitzki, 1995). EMAP also addresses the condition of landscape ecological conditions under EMAP – L (Environmental Monitoring and Assessment Program – Landscapes). This particular programme is developing indicators that can be measured from remotely sensed images (Cain *et al.*, 1997), primarily land cover maps derived from satellite images (US Environmental Protection Agency, 1994).

EMAP as a whole has a uniform, systematic sampling grid that covers the whole of the USA. The point frame is hierarchical, consisting of a nested series of grids at increasing densities. The reference-level density (the basis for regional sampling) consists of 12 500 sampling grid points located approximately 27.1 km apart (see Figure 7.1). The nested nature of the sampling frame allows different indicators, or indicators in different ecosystem categories, to be measured at different levels of resolution. Ecosystem distributions can be measured from satellites or aircraft at relatively low cost and can therefore be measured at a higher grid density, whereas field surveys are usually based on measurements taken at lower grid densities. Sampling density can also be adjusted according to the scale of variation of ecosystems; for example, geographically restricted types need more intensive sampling than widespread ecosystem categories (Messer *et al.*, 1991).

As a general rule, EMAP describes resources in an area centred around grid points in two 'tiers'. Sampling within a hexagonal area of 39.7 km^2 around each grid point results in one-sixteenth of the land area of the USA being sampled overall. For Tier 1 sampling, existing information is used, or information is collected using remote sensing (aerial photography or satellite imagery). This generates a Tier 1 probability sample of the resources of the USA from which regional estimates of areal extent of all landscape entities can be generated as well as regional estimates of more discrete features like lakes or wetlands. Tier 1 can then be sub-sampled at random to provide a 'Tier 2' sample and obtain measurements of resource condition (Novitzki, 1995). There are cases where specific resource types are distributed too sparsely for the standard approach to generate enough survey sites, in which case Tier 1 sampling can be intensified.

A vital aspect in the success of any large monitoring programme is the management of data, their interpretation and the reporting of results. All too often, ecological monitoring generates information that is inaccessible and therefore never used. As emphasised by Messer *et al.* (1991), 'Monitoring programs often live or die on the basis of their data management structure'. For a large-scale, integrated monitoring programme like EMAP, it is necessary to ensure that data are accessible from research centres throughout the country. Constant quality checks are necessary to ensure consistency of sampling and interpretation. At the same time, the end-users of the information generated by the programme will have highly variable interests and technical abilities, making it necessary to develop innovative approaches to the analysis, interpretation and display of results. This whole process is facilitated greatly by a GIS-based approach to information gathering, storage and display that is consistent and easily 'shared'. Mapping scale is very important as it influences the types of decision that can be made. For example, Table 7.3 summarises the hierarchy of ecological units used in a different regional ecosystem classification and mapping system in British Columbia (Ecosystems Working Group, 1995). The table also shows the main planning or decision-making

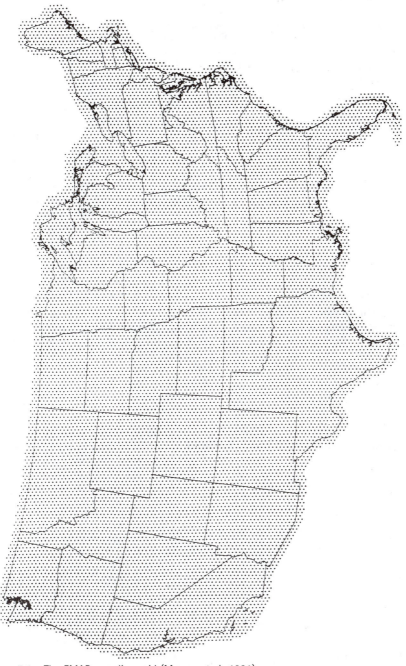

Figure 7.1   The EMAP sampling grid (Messer *et al.*, 1991).

Table 7.3 Ecological units used in ecosystem mapping in British Columbia (after Ecosystems Working Group, 1995)

Ecological unit	Common level of mapping	Common scale of presentation	Purpose, objectives and general use
Ecoregion units – stratification based on broad differences in climate			
Ecodomain	1 : 7 000 000	1 : 30 000 000	Area of broad climatic uniformity, relevant for international and national planning
Ecodivision	1 : 2 000 000	1 : 7 000 000	Area of broad climatic and physiographic uniformity, relevant for international and national planning
Ecoprovince	1 : 2 000 000	1 : 7 000 000	Area with consistent climate, used in provincial and regional planning
Ecoregion	1 : 100 000	1 : 2 000 000	Area with major physiographic and minor climatic variation
Ecosection	1 : 100 000	1 : 2 000 000	Area with minor physiographic and macroclimatic variation
Biogeoclimatic units – classes of ecosystem under the same regional climate: describe variations in vegetation and site conditions occurring within ecosections. The basic unit is the subzone, which may be grouped into zones or divided into variants or phases based on differences in regional climate (biogeoclimatic subzones and variants are the units mostly used in ecosystem mapping)			
Zone	1 : 250 000	1 : 2 000 000	Group of subzones: large geographical area with broadly homogeneous macroclimate, used in provincial and regional planning
Subzone	1 : 100 000	1 : 250 000	Climax or near-climax plant associations on zonal or typical sites, related to regional climate. Used in regional, operational and local planning
Variant	1 : 100 000	1 : 250 000	Reflects differences in average regional climate that result in climax plant subassociations on zonal sites
Phase	1 : 50 000	1 : 250 000	Reflects regional climatic variation resulting from local relief, e.g. extensive grasslands on steep south-facing slopes in an otherwise forested subzone
Ecosystem units			
Site series	1 : 5 000	1 : 20 000–1 : 50 000	Regional, operational and local planning
Site modifiers			
Structural stage			
Seral association			

applications of units mapped at different scales ranging from 1 : 5 000 to 1 : 30 000 000. For local decision making, more detailed information is generally required than for national or international planning applications, where a 'broad-brush' approach may suffice. The US Fish and Wildlife Service (1980) recommends use of maps generated from remotely sensed data at a scale of 1 : 20 000 to 1 : 60 000 to permit acceptable resolution for evaluating the distribution and availability of terrestrial wildlife habitat.

Measuring and mapping habitat potential

Some wildlife species are extremely difficult to census because of low population density and 'secretive lifestyles'. For such species, it can be more straightforward to infer possible distributions than to measure them directly. 'Potential distribution maps' may be derived by combining knowledge of relationships between species and habitat (derived from the literature, for example) with remotely sensed data on the distribution of landscape or ecological units. The usefulness of ecosystem maps for assessing habitat suitability depends on how well the attribute information stored in the GIS map database relates to key habitat requirements. Whereas grizzly bear denning habitat can be readily mapped at scales greater than 1 : 20 000, it is not distinguishable at a scale of 1 : 250 000. It is therefore necessary to consider the minimum size of map polygons needed to define meaningful habitat units in any particular case. Table 7.4 summarises the suitability of different mapping scales for deriving information about habitat use by animals. Smaller scales portray less detail regarding mapped ecosystem units and their attributes. Larger-scale maps provide much better detail over small areas, but for larger, more mobile species, may not provide the overview needed to assess overall habitat availability and quality.

The suitability of a map polygon for a species can be assessed in various ways, whether using professional experience, empirical data or modelled results of relationships between species' life requisites and ecosystem units.

Avery and Haines-Young (1990) used satellite imagery to map potential habitat for dunlin in the 'flow country' of Caithness and Sutherland, a remote part of northern Scotland where field surveys can be difficult and time-consuming. The near-infrared band 7 of Landsat Thematic Mapper images is sensitive to both vegetation type and ground wetness. Earlier field surveys in the area had suggested that breeding birds of this species would be most abundant in the wettest areas. Landsat TM imagery was used to predict numbers of breeding dunlin in randomly selected 2.5 × 2.5 km squares in the study area. These predictions were then checked using field data. There was a high correlation between the numbers predicted using the satellite imagery and the number counted 'on the ground' (Figure 7.2). This study demonstrated that it would have been possible to estimate populations of breeding dunlin over considerable (and remote) areas, based on a relationship established using limited field survey data. For further surveys, field counts would only have been required to verify predicted estimates.

Table 7.4   Ecosystem mapping scales appropriate to different levels of information about habitat use (after Wildlife Interpretations Sub-committee, 1996)

Scale of map	Information on habitat use
1 : 5000 ('very large')	Appropriate for species with restricted distributions or very specific habitat requirements (e.g. to identify suitable breeding ponds for great crested newts). Also useful for mapping interspersed habitats used by a large number of species – mapping at this scale targets small areas with high values (micro-habitats)
1 : 10 000–1 : 20 000 ('large')	Enables differentiation of ecosystem units and classification of level of use by study species – often used for land capability or land use planning, e.g. for forestry
1 : 50 000–1 : 100 000 ('medium')	Most useful for large vertebrate species that range over considerable areas and use several ecosystem units to meet their overall requirements – useful for relatively local or regional planning, where there is sufficient habitat knowledge to warrant detailed analysis at the landscape level
1 : 250 000–1 : 500 000 ('small')	Small map scales depict general ecological boundaries – generally used at regional or provincial planning levels. Approximate abundance of important habitat types can be quantified and used to assess relative status

Figure 7.2   Relationship between number of breeding dunlin predicted from satellite imagery and counts obtained using standard field survey methods: Predicted observations per cell: □ < 10, ▨ 10–30, ▦ 30–50, ■ > 50 (Avery and Haines-Young, 1990).

Table 7.5 Framework for predicting species distribution in landscape (Tucker *et al.*, 1997)

1. Collate the potential habitat preferences of target species from the scientific literature
2. Link these habitat preferences with the environmental data held in the GIS directly or by deriving 'surrogate habitat variables'. These are variables that can be readily extracted from the GIS, which are the nearest representation (in ecological terms) to the true habitat characteristics preferred by the species, as selected in stage 1
3. Assign conditional probabilities to the habitat variables held in the GIS, given the presence or absence of the species
4. Calculate the prior probability of the species being found anywhere in the landscape, irrespective of habitat
5. Assess the extent of independence of predictor habitat variables
6. Perturb the conditional probabilities associated with each GIS habitat variable over the range of possible joint probabilities if predictor variables are non-independent. Calculate the posterior probability of the species being found, given a particular suite of GIS variables at each point in the landscape, for each perturbation
7. Buffer around areas of habitat that are avoided by the species (e.g. urban areas, afforested areas). Remove from the map predicted areas of habitat that are too small to form viable territories
8. Output final GIS maps of predicted species distribution

For many species, it is not so easy to demonstrate a clear or predictable relationship between distribution and habitat variables that can be measured using remotely sensed data alone. Often a variety of data sources must be tapped into and interpreted before clear links can be established.

Tucker *et al.* (1997) used a combined GIS and Bayesian rule-based approach to model bird distributions in north-east England. They were particularly interested in predictive distribution models to anticipate the likely effects of changes in land use. They developed a habitat suitability model for predicting distributions of breeding bird species that linked detailed information on habitat preference for nesting with spatially referenced environmental information stored in a GIS database. Table 7.5 summarises the approach taken.

Models were applied to three bird species: the coal tit (*Parus ater* L.), the golden plover (*Pluvialis apricaria* L.) and the snipe (*Gallinago gallinago* L.). Bayes' theorem for conditional probability was used to modify initial estimates of the probability of encountering a species in a landscape, using the known preferences of the species for individual habitat characteristics and information on the distribution of those characteristics in the landscape. Figure 7.3 is a flowchart of the Bayesian–GIS modelling system used. Model complexity varied between species. For the coal tit, satellite land cover alone could be used as a habitat variable. For the snipe, satellite cover and altitude were necessary, while the golden plover model required satellite land cover and also information on altitude and gradient. However, the golden plover only breeds at gradients less than 10°, so this could be used as a filter and the three-tiered model effectively reduced to two tiers.

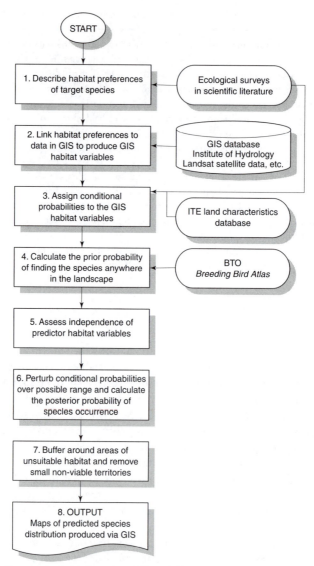

Figure 7.3   Flowchart of the Bayesian–GIS modelling system to model bird distributions in north-east England (Tucker *et al.*, 1997).

Measuring habitat distribution and re-distribution

Cumulative habitat loss has become a problem for many species in many countries. Not only has the amount of wildlife habitat declined progressively, but its distribution patterns have altered. It is well known that the geographic location and spatial organisation of habitat has a strong bearing on its accessibility and value to associated species. GIS technology has made it considerably easier to quantify those attributes of habitat distribution and organisation that might affect habitat value and therefore to recognise when unacceptable thresholds of habitat loss and fragmentation may have been reached for particular species. Attributes of habitat availability and organisation that might affect habitat 'value' and that are amenable to measurement using GIS include:

- amount or area of habitat (one or many types),
- habitat isolation (distances between habitat areas or 'patches' can be measured and compared with species' dispersal distances),
- edge : interior ratio (habitat may be more exposed to 'external influences', including pollution, disturbance and invasion by uncharacteristic species),
- habitat fragmentation (remaining habitat is in small, isolated units with higher edge : interior ratios), and
- average patch size (remaining habitat is in small units with higher edge : interior ratios – average patch size can be compared with minimum habitat area required to sustain different species).

Although 'areas' can be measured on any map, measurements of habitat area remaining or lost are considerably easier using GIS. Not only can land-take be quantified, but the implications of different extents and types of habitat loss can be compared. If information is available about the relationship between habitat area, habitat quality and carrying capacity (a big 'if'), it may be possible to predict the relative impacts of habitat loss for different species.

Treweek and Veitch (1996) used GIS and remotely sensed data to compare the effects of hypothetical new road routes on the amount and distribution of broadleaved woodland and lowland heathland in a 145 200 ha study area in the county of Dorset in England, mapped using the ITE's *Land Cover Map*. Local and regional impacts were quantified in terms of land-take and habitat fragmentation, both to compare the relative impacts of alternative routes and to assess the impacts of either with respect to existing or baseline conditions. Most importantly, use of GIS made it possible to study regional changes in land cover distribution in relation to the road network as a whole, rather than focusing on very local effects. However, two problems emerged during the course of this study. One was associated with the fact that the dimensions of linear features like roads were often exaggerated to make them more prominent or visible on maps. Where paper maps are digitized for incorporation into a GIS, as in this case, locational inaccuracies are perpetuated, making it difficult to determine exactly where roads are located with respect to habitat or to calculate the actual area of land (or a wildlife habitat) that would be occupied or destroyed. To get round this problem, an alternative and perhaps more 'honest' approach was tried, using data summarized at the 1 km^2 level to estimate the overall potential or risk for habitat loss or fragmentation to occur, but this meant working with an artificial grid that may have concealed important local variations in habitat distribution and resulted in imprecise estimates of actual impacts. This serves to emphasise that the usefulness of GIS-based studies is severely compromised if the data going in are flawed in any way. On the other hand, the same applies to any ecological study, how ever it is carried out.

Many authors have demonstrated the value of GIS as an important tool for assessing cumulative ecological effects at the landscape level, (for example, see papers by Cocklin *et al.*, 1992; Johnson *et al.*, 1988; Sebastini *et al.*, 1989; Walker *et al.*, 1986 and 1987). This is partly because of the ability to construct GIS to map or model ecological impacts expressed over large geographical scales, using remotely sensed data. The studies by Sebastini *et al.* (1989) and Johnson *et al.* (1988) demonstrate how the use of GIS can facilitate studies of cumulative ecological changes expressed over large areas or long time periods. Sebastini *et al.* (1989) used a GIS-based approach to trace the cumulative loss of coastal wetlands as a result of development in a large study region in the USA. Johnson *et al.* (1988) used GIS-based analysis to assess the effects of wetland loss on water quality and flood attenuation using land use maps spanning many years to quantify the magnitude, location and rate of wetland loss as a result of development and to project possible future rates of loss. The relationship between water quality and the extent and position of wetlands was then modelled using empirical data from 37 watersheds. Studies of this kind benefit from the ability to address spatial and temporal trends in habitat distribution and also the ability to study correlations between habitat presence and other features of interest, like water quality. Without GIS, such studies would be incredibly laborious, time-consuming and costly.

The application of such approaches demands:

- selection of a mapping scale that provides sufficiently detailed information about sources and receptors;
- availability of data on potential sources of cumulative impact – their spatial and temporal range and magnitude;
- local, ecological landscape mapping undertaken to fit within broader, regional mapping frameworks (e.g. to facilitate understanding of a project area's contribution to regional habitat supply and allow accurate assessment of regional project impacts); and
- use of a landscape classification and mapping system that captures both landscape features and ecosystem processes at both local and regional levels.

GIS as a tool in studying ecological processes: distribution of orchids in a sward

While several of the examples in this chapter have made use of GIS over large areas it is the amount of the data rather than how far apart the observations are that is important. As an example of GIS used over a very small area we include a study of the distribution of orchids within a sward. The Totternhoe site near Bedford (England) was used as a chalk quarry during the medieval period, but this activity ceased several hundred years ago leaving a pockmarked land surface of small pits and mounds with very steep (but short) slopes so that conditions can be markedly different over distances of a few centimetres. The balance between open grass and scrub is maintained by grazing and air photography shows marked fluctuations over decadal time scales. The site is important for the number and variety of terrestrial orchids, although the 'orchid-rich' part of the site is only 0.5 ha (roughly an acre). For 30 years the position of every flowering spike of two varieties of orchid has been recorded. The aim of the study was to determine how the micro-topography affects the flowering success of the different species and how different parts of the site are exploited from year to year. These data are currently being analysed to relate the success of individual plants to micro-topographic features under different periods of climatic conditions (Ruth Cox, *pers. comm.*).

GIS as a tool in conservation management: ITE's wetlands GIS

Nature conservation management in the more industrialised nations of the world is often primarily site-based. Even so, it is invariably necessary to take account of a wide range of factors in deriving suitable management strategies. For many wildlife habitats and species, the success of conservation on any particular site often depends on processes operating externally as well as those that can be managed internally. There are some examples where use of GIS has helped to explore multi-faceted conservation problems and to derive suitable management strategies.

In one such example, GIS was used to explore the likely implications and effectiveness of a strategy to restore lowland wet grassland habitats in the UK by raising the soil water table. The UK's Ministry of Agriculture, Fisheries and Food (MAFF) has designated a number of Environmentally Sensitive Areas (ESAs) in which farmers are offered incentives to manage their land in a manner consistent with achievement of certain environmental objectives. A number of ESAs have significant areas of lowland wet grassland, for which MAFF has prescribed agricultural and water-management operations intended to conserve or restore their distinctive wildlife (notably wading birds and certain characteristic vegetation communities). The implications of altering water levels depend on many interrelated factors that operate at a range of scales. Management prescriptions and financial incentives are uniform, rather than being tailored to suit different site conditions. Clearly, however, the implementation of uniform management on sites with very different soil types, management histories and current wildlife interest will have a variety of outcomes. To ensure prescribed management achieved an acceptable balance between ecological and agronomic objectives on different sites, evaluation of the effects of management was required. Without the benefit either of hindsight or of preparatory field research on the full range of potential sites, such evaluation was difficult. This case study shows how a GIS-based approach was used to model the outcomes of management on different sites and to resolve some of the conflicts of interest that can arise when attempting to manage for multiple land use objectives.

The issues listed in Box 7.1 illustrate the difficulties in prescribing management that would achieve both nature conservation and agronomic objectives on the same land (after Brown *et al.*, in press).

One of the reasons why a GIS-based approach was used to evaluate management prescriptions was the fact that effective hydrological management (either

Box 7.1   Selected nature conservation and agronomic issues to be taken into account when prescribing management

Nature conservation issues

Do current hydrological conditions pose a threat to important species or communities?

Which species are currently most at risk and in what way?

Are these species unique to this area or are they widespread?

Will restoration of wet grassland biodiversity demand hydrological change?

Will proposed hydrological management create appropriate hydrological conditions for target species or communities?

Which areas of land are most suitable for creation of appropriate hydrological conditions?

Of those areas that are suitable, which are available for restoration?

Agronomic issues

Which areas of land would be eligible for ESA 'tier payments' (incentive payments) with little intervention or management change?

If hydrological conditions change, what are the implications for agricultural production? For example, will increased soil wetness demand changes in livestock management, machinery use or grassland harvesting dates?

Will productive potential be lost and will ESA payments compensate for lost profit?

for nature conservation or for effective farming) was rarely possible on a 'field-by-field' basis (Treweek *et al.*, 1991). Not only can manipulation of water levels in one field have significant effects on hydrological conditions in adjacent fields, but water supplies are increasingly threatened by demands from domestic and industrial users. This means that the availability of water needed to achieve target 'wetness regimes' often depends on water management throughout whole catchments. Furthermore, for mobile organisms such as birds and mammals, a matrix of habitat types may be necessary to satisfy habitat requirements throughout the year (Forman and Godron, 1986). Therefore, while it might be possible to research the water requirements of individual species or communities through controlled experimentation in the laboratory or in the field, the results of such research cannot be applied without considering their spatial or 'landscape' context.

Management of land for multiple objectives (in this case agronomic, ecological and economic) required strategies based on understanding of the mechanisms determining the likely outcomes of different management options. The wetlands GIS was developed to:

- assess links between distributions of wetland communities, hydrological regime and agricultural management; and
- predict the likely implications of alternative water management actions for both agricultural and ecological interest within defined test catchments.

A GIS-based approach had particular advantages over manual mapping in the areas of data organisation and management as well as providing the opportunity for describing and modelling spatial relationships.

Information derived from field surveys was related to a base map. Colour aerial photography was used to interpret land use and produce digital boundaries that were imported directly into the GIS (a process that included ortho-correction of the aerial photography). Nine land use classes were interpreted from the aerial photography, with seven sub-classes: field boundaries, watercourses (stream/river, drainage channel), woodland (coniferous, broadleaved, mixed), scrubland, arable, grassland, ridge-and-furrow (arable, grassland), disturbed ground and wetland.

As the majority of the data were recorded at 'field level', field numbers were used as unique 'identifiers' to link information held in the relational database with maps. Data drawn from the results of the field survey included:

- plant species lists for individual fields (species presence/absence),
- mean percentage cover of 197 plant species in 220 fields,
- plant species distribution maps for the study area,
- vegetation community maps,
- bird counts for selected fields,
- dipwell data recording soil water levels for selected areas,
- measurements of soil penetrability (for feeding waders),
- estimates of soil invertebrate biomass (potential food supply for birds),
- sward height at the time of survey, and
- current agricultural land use and management.

To these field data, summary information on the known ecological requirements of particular species were added. For plants, for example, a system of indicator values developed by Ellenberg (1988) was used to create a database of moisture (F) and fertility (N) indicator values for all the species recorded in the study area. This could be used as a basis for predicting plant species composition in unsurveyed fields, given knowledge of soil wetness or soil nitrogen levels.

The use of the wetlands GIS is best illustrated through an example that explains how maps and data could be interrogated. A number of bird species of nature conservation importance occur in lowland wet grassland, including snipe and lapwing. The feeding behaviour and habitat requirements of these two species of wader differ. The wetlands GIS was used to establish which fields were used consistently by lapwing and snipe within the study area and then to determine whether the fields used consistently over the study period differed in terms of soil wetness or food availability?

Fields were classified as being used 'consistently' if birds were present in at least two winters during the study period of three years. By querying the database of field survey information, it was found that distribution maps for the two species tended to be mutually exclusive, though both species apparently favoured riverside fields. The reasons for these distributions could be investigated through the relational database containing the characteristics of those fields used by snipe and lapwing. Preliminary examination of the habitat preferences of snipe and lapwing in the study area suggested that snipe were more frequent in the wetter fields, which were also rich in invertebrates, whereas lapwing appeared able to tolerate fields with low invertebrate abundance (Treweek *et al.*, 1994; Caldow and Pearson, 1995). To examine these patterns further, recorded information on soil penetrability, soil wetness and invertebrate biomass was accessed and queried for the fields used by snipe.

Recorded values for soil penetrability ranged from 0.0 to 9.01 with an average of 4.9 (measured in KgF) (higher values indicate that greater force is necessary to penetrate the soil surface and probe for invertebrates as food). The fields favoured by lapwing, rather than snipe, were those with drier, less penetrable soils and lower invertebrate biomass. Soil penetrability may not be as important for lapwings because they are 'near-surface feeders' (reflected in their bill morphology), in contrast to snipe, which rely on an ability to probe the soil for invertebrate food supplies.

Differences between species distributions and preferences could be quantified by using a GIS-based approach that combined a relatively intensive field survey with the use of digitised aerial photography. Relationships between soil water table and species distribution could be explored, as well as the possible implications of altered soil wetness for agricultural land use. The results could be compared with those from related studies based on rigorous field survey and controlled laboratory experiments and also used to structure subsequent field surveys to best effect.

Note that effective communication of the results of such studies is vital if management decisions are to be improved. Many types of cartographic output can be generated from GIS, most frequently screen displays of the results of specific analyses. The output may also take the form of hard copy for use in reports, or be saved as digital map displays (such as screen dumps) for insertion into other computer display packages such as desk-top mapping packages. These enable the wider dissemination of results in a visual form that demonstrates key findings effectively and makes them more accessible.

When (not) to use GIS for ecological studies ——————

For ecological studies addressing wide-ranging effects in relatively inaccessible areas, the cost of field survey can be prohibitive. In these circumstances, using GIS opens up opportunities for using alternative source data (aerial photographs or satellite imagery) and facilitates the study of wider-scale impacts. On the other hand, Johnston (1998) rightly emphasises the tedious and time-consuming nature of map digitising: a task that should not be taken lightly. In one example, the average time taken to interpret aerial photographs, carry out field checks and draft inventory maps for wetlands was 58.3 h per 36 mile^2. However, in some situations, a GIS-based approach can be cheaper and more efficient than a traditional approach based on field inventory and hand-drafting of maps to present results. Most importantly, automated ecological study using GIS comes into its own as a basis for monitoring. Satellites can repeat land cover surveys much more frequently, efficiently and accurately than people on foot. When considering the use of GIS for ESA it is therefore important to ask whether or not GIS will make the process more efficient, cost-effective and productive than more traditional methods. It is worth noting that there are some ecological problems that were far too intractable to tackle before the advent of GIS. It is undoubtedly a technology that has opened up new avenues for research and analysis as well as new sources of data.

At present, the lack of readily available digital ecological data means that a certain amount of hand digitising is very likely to be necessary. This can be considered as a project 'overhead' that may, or may not, be worthwhile, depending on the use that is made of the data. If no complex interrelationships are involved, the cost of data entry may well outweigh the benefit of developing a GIS. If, on the other hand, complex relationships do need to be considered, particularly if iterative analyses are likely to be performed, the initial cost of data entry may well be justified as each layer of data is likely to be used more than once (McAuley, 1991). In terms of access to digital data for ecology in general, there are some initiatives that might enhance the availability of digital data. These initiatives have focused not only on the collation of data, but also on the provision of important information about the data that are collected, such as their source, date, scale, accuracy and reliability (Joao and Fonseca, 1996). They include the National Spatial Data Infrastructure (NSDI) in the USA (US Executive Office of the President, 1994), the National System for Geographical Information (SNIG) in Portugal (Henriques, 1996), the National Geospatial Database (NGD) in Great Britain (Nanson *et al.*, 1995) and the Geographic Information System of the Commission of the European Communities (GISCO), which was referred to earlier.

Common drawbacks when attempting to use GIS include:

- lack of availability of digital data (but see above),
- lack of time for data collection and entry,
- lack of experience and familiarity with software,
- false precision (obscuring sources of error),
- a technology-led approach, and
- over-investment in data irrelevant to decision making.

Johnston (1998) therefore gives the following advice to anyone considering a GIS-based approach:

1. Keep it simple.
2. Ask whether a GIS is necessary to tackle key questions.
3. Use existing data where possible rather than developing new databases.
4. Plan ahead: conceptualise, for example using data management systems or flowcharts to guide GIS development.
5. Keep good records, including a description of source data and analysis performed for each step in the GIS process.
6. Check results: is GIS output logical?
7. Consult with experienced individuals for advice on database management, data needs and GIS procedures.

L'envoi

We hope that we have given you a taste of the opportunities and pitfalls of using GIS to study ecological phenomena. Whether you are interested in the competition between grass and orchids in a sward or the range utilisation of grizzly bears a GIS can help in organising data and revealing new relationships and patterns. So now is the time to stop reading and start doing.

Good luck!

Appendix

Contacts for UK environmental/ecological data in electronic format

Data set	Contact	Additional information
Land Cover Map of Great Britain	Sue Wallis, ITE, Abbots Ripton, Huntingdon, Cambridge, PE17 2LS, UK Tel: +44(0)1487 773381 Fax: +44(0)1487 773467	Cost dependent on area required (staff time to extract and send the data), e.g.: 500 km² for a non-commercial application about £300 New version will eventually be available from data captured 1998 to 1999
Species data	Sue Wallis, ITE, Abbots Ripton, Huntingdon, Cambridge, PE17 2LS, UK Tel: +44(0)1487 773381 Fax: +44(0)1487 773467	Most data collected by dedicate volunteers: spatial and temporal coverage can vary
Digital terrain model	David Morris, IH, Maclean Building, Crowmarsh Gifford, Wallingford, Oxon., OX10 8BB, UK Tel: +44(0)1491 838800 Fax: +44(0)1491 692424	Cost subject to negotiation (need to persuade them it is for research) DTMs also available from the OS and the Ministry of Defence
Human census data	MIDAS, Manchester Computing Centre, The University of Manchester, Oxford Road, Manchester, M13 9PL, UK Tel: +44(0)161 275 6042 Fax: +44(0)161 275 6040	Free for academic research (but need to become a registered user) Extrapolation based on centroids of enumeration districts and are estimated by 'best judgement'
Precipitation, evaporation and river flow	David Morris, IH, Maclean Building, Crowmarsh Gifford, Wallingford, Oxon., OX10 8BB, UK Tel: +44(0)1491 838800 Fax: +44(0)1491 692424	Cost depends on use Meteorological variables are interpolated Time series may not be complete, accuracy of some gauging weirs limited at high flow rates

Data set	Contact	Additional information
Soils	Ian Bradley, Computing and Information Systems Department, Soil Survey and Land Research Centre, Cranfield University, Silsoe, Bedford, MK45 4DT, UK Tel: +44(0)1525 863259 Fax: +44(0)1525 863253	Soil association data – 100 m resolution £1200 for a 50 km² tile (subsequent years approximately 30% of first year's cost) For 1 km resolution – £600 for a 100 km² tile (subsequent years approximately 30% of first year's cost) HOST data – 1 km resolution, £250 per 50 km² tile – a key to allocating HOST classes to soil associations is £250
Solid geology	Kevin Becken, BGS, Keyworth, Nottingham, NG12 5GG, UK Tel: +44(0)115 936 3241 Fax: +44(0)115 936 3488	Electronic data flows – published map sheets covering 1° latitude by 2° longitude. Vector format data £600 for first year, £500 thereafter; raster format £450 for first year, £300 thereafter
Agricultural land use	Richard Reed, MAFF, Government Building, Epsom Road, Guildford, Surrey, GU1 2LD, UK Tel: +44(0)1483 68121 Fax: +44(0)1483 403976	Totals for the whole country for the last ten or more years (in spreadsheet format) can be downloaded directly from the Internet. Parish data costs open to negotiation! Only covers land in agricultural holdings that are classed as full-time

Glossary of terms

accuracy – deviation between a measured value and the true value

aliasing – (also called 'jaggies') occurrence of jagged lines on a computer screen or other raster display, caused by approximating a line to a series of cells – occurs where the level of detail exceeds the available resolution, most disconcerting when it happens to text

AM/FM – automated mapping/facilities mapping – term commonly used with reference to asset management of 'utilities', such as water, power and telecommunications companies

angular resolution – a measure of the smallest angular separation between two objects – the smallest object that can be seen

anisotropy – changes in a property is not independent of direction

arc – a line (does not have to be curved!) consisting of a series of **vertices** joined by straight sections

attribute accuracy – accuracy of the attribute

autocorrelation – (sometimes *spatial* autocorrelation) the degree to which values are influenced by the surrounding values; an essential property of phenomena that can be interpolated

axis – a reference line in a co-ordinate system

band – a defined interval in the electromagnetic spectrum (see also *panchromatic* and *multispectral*)

base map – background information used to provide 'context' for other layers of information, usually relatively permanent and timeless, such as topography, soils, geology, but may be more dynamic, such as administrative boundaries

CAD – computer-aided design/drafting (not really a GIS term, but is often encountered)

cadastral survey – public record of the extent and ownership of land, from the French *cadastre* (sometimes also called metes and bounds surveys)

Cartesian co-ordinates – a method of representing location using values along two (or three) axes perpendicular to each other (first devised by René Descartes, 1596–1650). Also sometimes called *rectangular co-ordinates*

cartogram – a diagram (rarely used in ecology) where space is deliberately distorted to allow comparison of values between regions with very different spatial extent; the skill is in allowing the user to recognise the regions after the distortion has been applied

cartography – the art of portraying some aspects of the real world as graphical features on a two-dimensional surface. *Digital* or *automatic* cartography makes use of computers and software to help in the process

167

change detection – comparison of two images of the same area acquired at different times (a surprisingly difficult operation using satellite data)

choropleth – a map where phenomena are represented by homogeneous polygons

clip – extraction or display of features within a polygon

completeness – whether the digital data cover all the relationships and objects that they are supposed to

contour – line connecting points of equal elevation – a particular type of isoline

conversion – process of transferring data into a digital database – this is *the* major input problem in GIS and is often very time-consuming

convex hull – the shortest line around a series of points

covariance – the degree with which points close together tend to be similar

covariogram – (also covariance function) the relationship between the *covariance* and the separation of pairs of points. If Y(s) is a spatial stochastic process then the covariance can be estimated as $C(h) = 1/n(h)\Sigma(y_i-y)(y_j-y)$, where $n(h)$ is the number of pairs of points that are h distance apart

coverage – a set of coherent information about a particular theme, also termed a layer

currency – the degree to which data conform to the time period they are supposed to

cursor – see *puck*

dangle – the most common error in digitising, where lines cross instead of meeting; needs to be corrected by *snapping* the *nodes* together

datum – usually a reference plane when measuring elevation

digitising – a manual method for converting graphical information on paper into digital data (see also *scanning*). The resultant data are in the form of points and lines (*vector* data)

dpi – dots per inch – a measure of resolution – the higher the dpi the finer the resolution

extrapolation – prediction made beyond the area covered by observations – not to be recommended but sometimes inescapable (see also *interpolation*)

false origin – the true origin is in the centre of the area to be mapped (to minimise the average distortion) but that means that there will be positive and negative co-ordinates. To avoid this, large numbers are added to the *x*- and *y*-co-ordinates to create a false origin outside the area of interest

geographic co-ordinates – method of specifying location in terms of angles measured from the centre of the Earth i.e. *latitude* and *longitude*

geographic information system (GIS) – a collection of hardware and software for dealing with spatial information – also the process of manipulating spatial data

geoid – an irregular equipotential surface closely approximating to global mean sea level (see also *spheroid*)

georeferencing – locating a feature within a model of the surface of the Earth – subtly different to *geocoding*: the process of associating a geographic location to non-geographic data

global accuracy – accuracy relative to an absolute frame of reference (also absolute accuracy)

global positioning system (GPS) – a method of locating the position of a receiver by measuring the distances from a swarm of satellites

graphical user interface (GUI) – system that allows a user to interact through a mouse (or other pointing device) as well as a keyboard

graphical kernel system (GKS) – a series of specifications that allow graphics to be drawn using Fortran programs

graticule – lines printed on a map to aid a user to locate particular co-ordinates

hardware – the physical parts of a computer system: central processing unit (CPU), visual display unit (VDU), keyboard, mouse, scanner, digitisers, printers, plotters, hard disks, CD-ROMs, optical disks, tape streamers and networks

heteroscedasticity – variance of errors is not constant over the whole region

hue – one of the aspects of colour, dependent on the dominant wavelength

hypsometric – refers to a sequence of colours used to represent elevation

image enhancement – improving the visual appearance of an image

image processing – turning the image into information

information – often used as a synonym for data, but really should be used for data that have been interpreted or have some other 'added value'

interpolation – making predictions within the area of observations (see also *extrapolate*)

isoline – a line joining equal values

isopleth – map consisting of lines (*isolines*) joining equal values

isotropy – property being investigated is independent of direction

land information system (LIS) – a GIS-like construct often used in conjunction with *cadstral* or *AM/FM* mapping

legend – a written description of the symbols used

liveware – you! – the person who provides the motive for doing the analysis and controls the hardware and software

local – accuracy between neighbouring points (also relative accuracy)

logical consistency – whether the representation of objects (within the GIS) is correct; for example, if a river is represented as a network of lines do the correct lines join at the confluences

metadata – information about data, such as who collected it, when, why, precision, accuracy, etc.

multispectral – measuring the scene over a number of distinct narrow *bands* in the electromagnetic spectrum – typical sensors use three or four *bands* but can exceed a hundred

neatline – the border around a map, title, key, etc.

node – the end points of a line (special case of a *vertex*) – if two lines cross but do not share a node they will not interact

origin – the point from which all co-ordinate positions are referred to (the point with co-ordinates 0,0)

orthophoto – a geometrically corrected air photography used to form a backdrop of a map

orthorectification – process of geometrically correcting an air photograph

panchromatic – measuring the reflectance over a wide *band* of the spectrum; typically the range will be close to that seen by the human eye

photogrammetry – process of extracting geometric (shape, size) information from photographs

pixel – contraction of 'picture element' – in remote sensing a pixel has spatial and spectral properties but is often used more loosely in GIS to indicate an individual cell in a raster grid or matrix

polygon – one or more *lines* or *arcs* forming a closed loop (none of the *arcs* forming a GIS polygon has to be straight)

positional accuracy – accuracy of the location of a point, line, etc.

precision – amount of 'detail' or number of significant digits in a measurement – a distance measured to the nearest millimetre is more precise than one measured to the nearest centimetre (but not necessarily more accurate)

projection – a series of equations to convert co-ordinates in latitude–longitude to eastings–northings (or *x–y*) and vice versa

puck – the pointing device on a digitising *tablet* (roughly equivalent to a mouse on a computer) – also called a *cursor*. The cross hairs on the puck are lined up on the point to be captured, and when a button on the puck is pressed the location on the tablet is recorded

quadtree – a method of storing data where the resolution is high (small cells) where there is a lot of detail and low (large cells) where there is little detail

raster – a method of storing information in a regular grid or matrix of points – location is inferred from the cell size and the boundaries of the grid – a raster GIS uses grids as the primary means of representing information

relational database – a database where data in a row or column in a table are related to data in a row or column in another table

rectification – process of aligning an image to a particular map projection and removing distortions due to the imaging system, viewing angle

reflectance – ratio of the energy reflected by a body to that incident on it – depends on wavelength and direction

registration – process of aligning the image with another image or map

relational data base management system (RDBMS) – the most common form of database associated with a GIS – data are organised as a series of tables that are related together (also shortened to RDB)

resolution – smallest difference that can be distinguished

saturation – how intense a colour is

scale – the relationship between the actual distance between two features and the separation between the representation of the two features on the map

scanner – machine for converting graphical data, e.g. printed maps, into digital data

scanning – an automated method for converting graphical information on paper into digital data (see also *digitising*) – the resultant data are in the form of *raster* grids

sliver polygons – thin polygons that result when overlaying polygons due to slight inconsistencies in how the boundaries were specified

snapping – process of correcting digitising errors by making lines or points that should join 'snap' together – it may be necessary to insert another node first

software – programmes and instructions that manipulate data and information using the computer *hardware*

spatial analysis – in early text used to mean spatial statistics – more recently (and in this text) used more loosely to include spatial search, map overlay, network analysis and other processes that do not involve testing statistical inferences

spectral signature – a unique pattern of reflectance resulting from incident energy with a range of wavelengths – central to the idea of distinguishing different objects from how they reflect or absorb electromagnetic radiation

spheroid – regular approximation to the *geoid* – typically an ellipsoid (an ellipse rotated around the minor axis)

stationarity – arbitrarily selected sub-regions will have similar values for the same property (mean and variance are independent of location)

stereo imaging – two images of the same area taken from slightly different points so that the illusion of a three-dimensional object can be recreated

stereoscopic – refers to two slightly different views of the same subject – used to create a three-dimensional view

tablet – a table that has many fine wires embedded in it for detecting the magnetic field generated by the ***puck*** (and so locating the position)

thematic map – map with a particular theme (one of the great mysteries of ***cartography*** is, how could you have a non-thematic map: a map about nothing?) – sometimes called a *data map*

theodolite – surveying instrument for measuring angles in the vertical and horizontal planes

topology – the branch of mathematics concerned with the spatial relationship between features

track – or ground track–vertical projection of the flight path

triangulation – locating a position by measuring angles

trilateration – locating a position by measuring distances

traverse – locating a position by measuring distances and angles

variogram – (strictly semi-variogram) the relationship between the variance between points and their separation – can be estimated as $2\gamma(h) = 1/n(h)\Sigma(y_i-y_j)^2$, where $n(h)$ is the number of pairs of points h distance apart

vector – in GIS, a line used to represent some feature or property – a vector may directly represent linear features or enclose the object or feature to form polygons – a 'vector GIS' uses lines and points as the primary means of representing information

vertex – a point on a line with known co-ordinates

vertices – plural of ***vertex***

References

Albrecht, J., 1996, *Universal GIS Operators – A Task-Orientated Systematization of Data Structure – Independent GIS Functionality Leading Towards a Geographic Modeling Language,* Technical Report, ISPA Vechta Germany.

Anon., 1989, The case against rectangular world maps, *The Cartographic Journal,* **26**, 156–157.

Avery, M.I. and Haines-Young, R.H., 1990, Population estimates for the dunlin *Calidris alpina* derived from remotely sensed satellite imagery of the Flow Country of northern Scotland, *Nature,* **344**, 860–862.

Bailey, T.C. and Gatrell, A.C., 1995, *Interactive Spatial Data Analysis,* Harlow: Longman.

Bina, O., Briggs, B. and Bunting, G., 1997, Towards an assessment of trans-European transport networks' impact on nature conservation. *Proceedings of the International Conference on Habitat Fragmentation and Infrastructure,* edited by K. Canters. Dutch Ministry of Transport, Public Works and Water Management, Delft.

Bolstad, P.V. and Smith, J.L., 1995, Errors in GIS: assessing spatial data accuracy, pp. 301–312, in Lyon, J.G. and McCarthy, J. (eds), *Wetland and Environmental Applications of GIS,* New York: Lewis.

Brewer, C.A., 1996, Guideline for selecting colours for diverging schemes on maps, *The Cartographic Journal,* **33**, 79–86.

Brown, N.J., Swetnam, R.D., Treweek, J.R., Mountford, J.O., Caldow, R.W.G., Manchester, S.J., Stamp, T.R., Gowing, D.J.G., Solomon, D.R. and Armstrong, A.C., in press, Issues in GIS development: adapting to research and policy needs for management of wet grassland in an environmentally sensitive area, *Journal of Environmental Management.*

Burrough, P.A., 1986, *Principles of Geographic Information Systems for Land Resource Assessment,* Monographs on Soil Resources Survey No. 12, Oxford: Oxford Scientific.

Cain, D.H., Ritters, K. and Orvis, K., 1997, A multi-scale analysis of landscape statistics, *Landscape Ecology,* **12**, 199–212.

Caldow, R.W.G. and Pearson, B., 1995, *The conservation and enhancement of biological diversity in farmland management: the relationship between environmental characteristics and the distribution of wetland birds.* ITE Report to the Ministry of Agriculture, Fisheries and Food.

Chambers, R., 1983, *Rural Development: Putting the Last First,* Harlow: Longman.

Cocklin, C., Parker, S. and Hay, J., 1992, Notes on cumulative environmental change. II: A contribution to methodology, *Journal of Environmental Management,* **35**, 51–67.

Coppock, J.T. and Rhind, D.W., 1991, The history of GIS, in Maguire, D.J., Goodchild, M.F. and Rhind, D.W. (eds), *Geographical Information Systems,* Vol. 1, Harlow: Longman.

Cox, R. and Parr, T.W., 1994, *Two Way Comparison of Land Cover Classifications*, Cambridge: Institute of Terrestrial Ecology.

Danson, M. and Plummer, S.B., 1995, *Advances in Environmental Remote Sensing*, London: John Wiley.

Dorling, D. and Fairburn, D., 1997, *Mapping Ways of Representing the World*, Harlow: Longman.

Ecosystems Working Group, 1995, *Standards for Terrestrial Ecosystems Mapping in British Columbia*. Ecosystems Working Group of the Terrestrial Ecosystems Task Force Resources Inventory Committee. Review Draft, 31 March 1995.

Edwards, T.C., Moisan, G.G. and Cutler, D.R., 1998, Assessing map accuracy in a remotely sensed ecoregion scale cover map, *Remote Sensing Environment*, **63**, 73–83.

Egenhofer, M.S. and Frank, A.U., 1987, Object oriented databases: databases requirements for GIS, pp. 189–211, in Aangeebrug, R.T. and Schiffman, Y.M. (eds), *International GIS Symposium*, Arlington, Virginia. Fall Church, Virginia: ASPRS.

Ellenberg, H., 1988, *Vegetation Ecology of Central Europe*. Cambridge: Cambridge University Press.

Food and Agriculture Organisation (FAO), 1997, *Africover Land Cover Classification*, Environment and Natural Resource Service Research, Extension and Training Division. FAO Sustainable Development Department, Rome: FAO.

Forman, T.T. and Godron, M., 1986, *Landscape Ecology*, New York: John Wiley & Sons.

Fuller, R.M., Groom, G.B. and Jones, A.R., 1994, The land cover map of Great Britain: an automated classification of Landsat Thematic Mapper data, *Photogrammetric Engineering and Remote Sensing*, **60**, 553–562.

Fuller, R.M. and Parsell, R.J., 1990, Classification of TM imagery in the study of land use in lowland Britain: practical considerations for operational use, *International Journal of Remote Sensing*, **11**, 1901–1917.

Gardingen van, P.R., Foody, G.M. and Curran, P.J., 1997, *Scaling-Up From Cell to Landscape*, Cambridge: Cambridge University Press.

GIS World Inc., 1991, *International GIS Sourcebook*, Fort Collins, Colo.: GIS World Inc.

Goodchild, M.F., 1990, Spatial information science, pp. 3–12, in *Proceedings of the Fourth International Symposium of Spatial Data Handling*, Vol. 1, Columbus, Ohio: International Geographic Union.

Haynes, K.E. and Fotheringham, A.S., 1984, *Gravity and Spatial Interaction Models*, London: Sage.

Henriques, R.G., 1996, The Portuguese National Network of Geographical Information (SNIG Network). *Proceedings of the Joint European Conference* (JEC) 27–29 March 1996, Barcelona.

Huxhold, W.E. and Levinsohn, A.G., 1995, *Managing GIS Projects*, Oxford: Oxford University Press.

Isaaks, E.H. and Srivastava, R.M., 1989, *An Introduction to Applied Geostatistics*, Oxford: Oxford University Press.

Joao, E. and Fonseca, A., 1996, *Current Use of Geographical Information Systems for Environmental Assessment: A Discussion Document*, Research papers in environmental and spatial analysis No. 36, London: London School of Economics.

Johnston, C.A., 1990, GIS: More than just a pretty face, *Landscape Ecology*, **4**, 3–4.

Johnston, C.A., 1998, *GIS: in Ecology*, Oxford: Blackwell.

Johnston, C.A., Detenbeck, N.E., Bonde, J.P. and Neimi, G.J., 1988, Geographical information systems for cumulative impact assessment, *Photogrammetric Engineering and Remote Sensing*, **54**, 1609–1615.

Kasischke, E.S., Melack, J.M. and Dobsom, M.C., 1997, The use of imaging radars for ecological applications – a review, *Remote Sensing Environment*, **59**, 141–156.

Keates, J.S., 1973, *Cartographic Design and Production*, Harlow: Longman.

Keates, J.S., 1982, *Understanding Maps*, Harlow: Longman.

Kilford, W.K., 1970, *Elementary Air Survey*, London: Pitman.

Krzanowski, W.J., 1988, *Principles of Multivariate Statistics: A User's Perspective*, Oxford: Oxford University Press.

Larner, A., 1992, Digital maps: what you see is not always what you get, in Cadoux-Hudson, J. and Heywood, I. (eds), *Geographic Information 1992/3 The Yearbook of the Association of Geographic Information*, London: Taylor & Francis.

Lillesand, T.M. and Keifer, R.W., 1987, *Remote Sensing and Image Interpretation*, New York: Wiley.

Maling, D.H., 1972, *Co-ordinate Systems and Map Projections*, London: George Philip & Son.

Mapping Awareness, 1997, Software survey, *Mapping Awareness*, **11**, 30–35.

McAuley, I., 1991, Environmental impact analysis: a cost-effective GIS application? *Mapping Awareness*, **5**(4), 36–40.

McGarigal, K. and Marks, B.J., 1994, FRAGSTATS version 2 documentation (downloaded from the Internet)

McHarg, I.L., 1971, *Design with Nature*, New York: Doubleday.

Messer, J.J., Linthurst, R.A. and Overton, W.S., 1991, An EPA program for monitoring ecological status and trends, *Environmental Monitoring and Assessment*, **17**, 67–78.

Monmonier, M., 1991, *How to Lie with Maps*, Chicago: University of Chicago Press.

Nanson, B., Smith, N. and Davey, A., 1995, What is the British National Geospatial Database? *AGI'95 Conference Proceedings*, Birmingham, pp. 111–145. London: The Association for Geographic Information.

Novitzki, R.P., 1995, EMAP-wetlands: a sampling design with global application, *Vegatatio*, **118**, 171–184.

Openshaw, S., 1984, *The Modifiable Areal Unit Problem, CATMOG 38*, Norwich: GeoBooks.

Pathirana, S., 1990, Fuzzy membership approach to the mixed pixel problem of remotely sensed data: an application in the suburban fringe zone of Northeast Ohio, PhD thesis, Kent State University.

Peterken, G.F. and Backmeroff, C., 1988, *Long-term Monitoring in Unmanaged Woodland Nature Reserves*, Research Report No. 9, Peterborough: English Nature.

Pipes, S., 1997, Countryside information system, *Mapping Awareness*, **11**(10), 28–29.

Plumb, K., Harris, R. and Needham, D., 1997, *Commercial Activities in High Resolution Imaging from Space*, 'Reference paper' BNSC Seminar 'From Space to Database' 26 February, London: Department of Trade and Industry.

Ramsey III, E.W. and Jensen, J.R., 1995, Modelling mangrove canopy reflectance using a light interaction model and an optimisation technique, pp. 61–81, in Lyon, J.G. and McCarthy, J. (eds), *Wetland and Environmental Applications of GIS*, New York: Lewis.

Reyment, R. and Joreskog, K.G., 1993, *Applied Factor Analysis in the Natural Sciences*, Cambridge: Cambridge University Press.

Robinson, J.M. and Zubrow, E., 1997, Restoring continuity: exploration of techniques for reconstructing the spatial distribution underlying polygonized data, *International Journal of Geographic Information Science*, **11**, 633–648.

Soil Survey Staff, 1983, *National Soils Handbook, Issue 1, Soil Conservation Service*, Washington DC: USDA.

Sebastini, M., Sambrano, A., Villamizar, A. and Villalba, C., 1989, Cumulative impact and sequential geographical analysis as tools for land use planning: a case study: Laguna La Reina, Miranda State, Venezuela, *Journal of Environmental Management*, **29**, 237–248.

Stamp, L.D., 1962, *The Land of Britain: Its Use and Misuse*, 3rd edition, Harlow: Longman.

Steers, J.A., 1927, *An Introduction to the Study of Map Projections*, London: University of London Press.

Terry, N.G., 1997, Field validation of the UTM gridded map, *Photogrammetric Engineering and Remote Sensing*, **63**, 381–383.

Treweek, J.R., Caldow, R.W.G., Armstrong, A.C., Dwyer, J. and Sheail, J., 1991, *Wetland Restoration: Techniques for an Integrated Approach*, ITE Report to the Ministry of Agriculture Fisheries and Food, Monks Wood: Institute of Terrestrial Ecology.

Treweek, J.R., Hankard, P. Roy, D.B., Arnold, H. and Thompson, S. (in press) Scope for strategic ecological assessment of trunk road development in England with respect to potential impacts on lowland heathland, the Dartford warbler (*Sylvia undata*) and the sand lizard (*Lacerta agilis*). *Journal of Environmental Management*.

Treweek, J.R., Manchester, S.J., Mountford, J.O., Sparks, T.H., Stewart, A., Veitch, N., Caldow, R.W.G. and Pearson, B., 1994, *Effects of managing water-levels to maintain or enhance ecological diversity within discrete catchments*. First ITE report to the Ministry of Agriculture.

Treweek, J.R. and Veitch, N., 1996, The potential application of GIS and remotely sensed data to the ecological assessment of proposed new road schemes, *Global Ecology and Biogeography Letters*, **5**, 249–257.

Tucker, K., Rushton, S.P., Sanderson, R.A., Martin, E.B. and Blaiklock, J., 1997, Modelling bird distributions – a combined GIS and Bayesian rule-based approach, *Landscape Ecology*, **12**(2), 77–93.

US Environmental Protection Agency, 1994, *Landscape Monitoring and Assessment Research Plan*, US EPA 620/R-94–009, Washington DC: Office of Research and Development.

US Executive Office of the President, 1994, *Coordinating geographic data acquisition and access: the National Spatial Data Infrastructure*. (Executive Order 12906). Executive Office of the President, Washington.

US Fish and Wildlife Service, 1980, *Habitat Evaluation Procedure* (HEP), ESM 102 Division of Ecological Services, Washington DC: US Fish and Wildlife Service, Department of the Interior.

Veitch, N., Treweek, J.R. and Fuller, R.M., 1995, The land cover map of Great Britain: a new data source for environmental planning and management, pp. 157–170, in Danson, F.M. and Plummer, S.E. (eds), *Advances in Environmental Remote Sensing*, Chichester: Wiley.

Vieux, B.E., 1995, Aggregation and smoothing effects on surface runoff modelling, pp. 205–229, in Lyon, J.G. and McCarthy. J. (eds), *Wetland and Environmental Applications of GIS*, New York: Lewis.

Wadsworth, R.A., Swetnam, R.D. and Willis, S.G., 1997, Seeds and sediment: modeling the spread of *Impatiens glandulifera* Royle, pp. 53–60, in Cooper, A. and Power, J. (eds), *Species Dispersal and Land Use Processes*, Aberdeen: International Association for Landscape Ecology.

Walker, D.A., Webber, P.J., Binnian, E.F., Everett, K.R., Lederer, N.D., Nordstrand, E.A. and Walker, M.D., 1987, Cumulative effects of oil fields in northern Alaskan landscapes, *Science*, **238**, 757–761.

Walkerm D.A., Webber, P.J., Walker, M.S., Lederer, N.D., Meehan, R.H. and Nordstrand, E.A., 1986, Use of geobotanical maps and automated mapping techniques to examine cumulative impacts in the Prudhoe Bay oilfield Alaska, *Environmental Conservation*, **13**, 149–160.

Welch, K. and Homsey, A., 1997, Datum shifts for UTM co-ordinates, *Photogrammetric Engineering and Remote Sensing*, **63**, 371–375.

Wickham, J.D., O'Neill, R.V., Ritters, K.H., Wade, T.G. and Jones, K.B., 1997, Sensitivity of selected landscape pattern metrics to land cover misclassification and differences in land cover composition, *Photogrammetric Engineering and Remote Sensing*, **63**, 397–402.

Wiens, J.A., 1989, Spatial scaling in ecology, *Functional Ecology*, **3**, 385–397.

Wildlife Interpretations Subcommittee, 1996, *Standards for wildlife habitat capability/suitability ratings in British Columbia*. Review Draft 20 August, 1996. Wildlife Interpretations Subcommittee, Ministry of Environment, Lands and Parks, Victoria.

Worboys, M.F., 1994, Object-oriented approaches to geo-referenced information, International Journal of GIS, **8**, 385–399.

Worboys, M.F., Hearnshaw, H.M. and Maguire, D.J., 1990, Object orientated data modelling for spatial databases, *International Journal of GIS*, **4**, 369–383.

Wu, H., Malafant, K.W.J., Pendridge, L.K., Sharpe, P.J.H. and Walker, J., 1987, Simulation of two-dimensional point patterns: application of a lattice framework approach, *Ecological Modelling*, **38**, 299–308.

Wyatt, B.K., Billington, C., Biede, K., Leeuw de, J., Greatorex Davis, N. and Luxmoore, R., 1997a, *Guidelines for Land Use and Land Cover Descriptions and Classifications*, final report to FAO (UN contracts FP/1003-94-52-2201 and 2202), Cambridge: Institute of Terrestrial Ecology.

Wyatt, B.K., Gerard, F.F. and Fuller, R.M., 1997b, *Correspondence to Other Themes as a Basis of Integrated Approaches*, report to the European Environment Agency, Cambridge: Institute of Terrestrial Ecology.

Index

(References to figures, tables, boxes and the glossary are indicated by the letters f, t, b and g respectively.)